VOLUME 502   MARCH 1989

THE ANNALS

of The American Academy *of* Political *and* Social Science

RICHARD D. LAMBERT, *Editor*

ALAN W. HESTON, *Associate Editor*

UNIVERSITIES AND THE MILITARY

Special Editor of this Volume

DAVID A. WILSON

Professor of Political Science
University of California
Los Angeles

$\textcircled{S}$ SAGE PUBLICATIONS   *NEWBURY PARK   LONDON   NEW DELHI*

THE ANNALS

© 1989 *by* The American Academy *of* Political *and* Social Science

ERICA GINSBURG, *Assistant Editor*

Editorial Office: 3937 Chestnut Street, Philadelphia, Pennsylvania 19104.

For information about membership* (individuals only) and subscriptions (institutions), address:

SAGE PUBLICATIONS, INC.
2111 West Hillcrest Drive
Newbury Park, CA 91320

From India and South Asia,
write to:
SAGE PUBLICATIONS INDIA Pvt. Ltd.
P.O. Box 4215
New Delhi 110 048
INDIA

From the UK, Europe, the Middle
East and Africa, write to:
SAGE PUBLICATIONS LTD
28 Banner Street
London EC1Y 8QE
ENGLAND

SAGE Production Editors: KITTY BEDNAR and LIANN LECH
* *Please note that members of The Academy receive THE ANNALS with their membership.*

Library of Congress Catalog Card Number 88-061072
International Standard Serial Number ISSN 0002-7162
International Standard Book Number ISBN 0-8039-3176-X (Vol. 502, 1989 paper)
International Standard Book Number ISBN 0-8039-3175-1 (Vol. 502, 1989 cloth)
Manufactured in the United States of America. First printing, March 1989.

The articles appearing in THE ANNALS are indexed in *Book Review Index; Public Affairs Information Service Bulletin; Social Sciences Index; Monthly Periodical Index; Current Contents; Behavioral, Social, Management Sciences;* and *Combined Retrospective Index Sets.* They are also abstracted and indexed in *ABC Pol Sci, Historical Abstracts, Human Resources Abstracts, Social Sciences Citation Index, United States Political Science Documents, Social Work Research & Abstracts, Peace Research Reviews, Sage Urban Studies Abstracts, International Political Science Abstracts, America: History and Life,* and/or *Family Resources Database. They are available on microfilm from University Microfilms, Ann Arbor, Michigan.*

Information about membership rates, institutional subscriptions, and back issue prices may be found on the facing page.

Advertising. Current rates and specifications may be obtained by writing to THE ANNALS Advertising and Promotion Manager at the Newbury Park office (address above).

Claims. Claims for undelivered copies must be made no later than three months following month of publication. The publisher will supply missing copies when losses have been sustained in transit and when the reserve stock will permit.

Change of Address. Six weeks' advance notice must be given when notifying of change of address to insure proper identification. Please specify name of journal. Send change of address to: THE ANNALS, c/o Sage Publications, Inc., 2111 West Hillcrest Drive, Newbury Park, CA 91320.

The American Academy of Political and Social Science

3937 Chestnut Street   Philadelphia, Pennsylvania 19104

Origin and Purpose. The Academy was organized December 14, 1889, to promote the progress of political and social science, especially through publications and meetings. The Academy does not take sides in controverted questions, but seeks to gather and present reliable information to assist the public in forming an intelligent and accurate judgment.

Meetings. The Academy holds an annual meeting in the spring extending over two days.

Publications. THE ANNALS is the bimonthly publication of The Academy. Each issue contains articles on some prominent social or political problem, written at the invitation of the editors. Also, monographs are published from time to time, numbers of which are distributed to pertinent professional organizations. These volumes constitute important reference works on the topics with which they deal, and they are extensively cited by authorities throughout the United States and abroad. The papers presented at the meetings of The Academy are included in THE ANNALS.

Membership. Each member of The Academy receives THE ANNALS and may attend the meetings of The Academy. Membership is open only to individuals. Annual dues: $30.00 for the regular paperbound edition (clothbound, $45.00). Add $9.00 per year for membership outside the U.S.A. Members may also purchase single issues of THE ANNALS for $7.95 each (clothbound, $12.00).

Subscriptions. THE ANNALS (ISSN 0002-7162) is published six times annually—in January, March, May, July, September, and November. Institutions may subscribe to THE ANNALS at the annual rate: $66.00 (clothbound, $84.00). Add $9.00 per year for subscriptions outside the U.S.A. Institutional rates for single issues: $12.00 each (clothbound, $17.00).

Second class postage paid at Philadelphia, Pennsylvania, and at additional mailing offices.

Single issues of THE ANNALS may be obtained by individuals who are not members of The Academy for $8.95 each (clothbound, $17.00). Single issues of THE ANNALS have proven to be excellent supplementary texts for classroom use. Direct inquiries regarding adoptions to THE ANNALS c/o Sage Publications (address below).

All correspondence concerning membership in The Academy, dues renewals, inquiries about membership status, and/or purchase of single issues of THE ANNALS should be sent to THE ANNALS c/o Sage Publications, Inc., 2111 West Hillcrest Drive, Newbury Park, CA 91320. *Please note that orders under $25 must be prepaid.* Sage affiliates in London and India will assist institutional subscribers abroad with regard to orders, claims, and inquiries for both subscriptions and single issues.

THE NINETY-FIRST ANNUAL MEETING OF THE AMERICAN ACADEMY OF POLITICAL AND SOCIAL SCIENCE

APRIL 28 and 29, 1989
THE BARCLAY HOTEL
PHILADELPHIA, PENNSYLVANIA

The annual meeting of The Academy is attended by many distinguished scholars, statesmen, authors, and professionals in diverse fields, including representatives of many embassies, academic institutions, and cultural, civic, and scientific organizations.

This 91st Annual Meeting will be addressed at each session by prominent scholars and officials and will be devoted to the topic of

HUMAN RIGHTS AROUND THE WORLD

Members of the Academy are cordially invited to attend and will receive full information. Information on Academy membership can be found in each volume of THE ANNALS.

- Proceedings of the 91st Annual Meeting will be published in the November 1989 volume of THE ANNALS.

- All members and attendees who have published a book may participate in the exhibits at the hotel. Contact Harve C. Horowitz & Associates, 11620 Vixens Path, Ellicott City, MD 21043, tel. (301) 997-0763.

FOR DETAILS ABOUT THE ANNUAL MEETING WRITE TO
THE AMERICAN ACADEMY OF POLITICAL AND SOCIAL SCIENCE
BUSINESS OFFICE ● 3937 CHESTNUT STREET
PHILADELPHIA, PENNSYLVANIA 19104

CONTENTS

PREFACE .... *David A. Wilson* 9

THE CONTEXT

THE U.S. MILITARY AND HIGHER EDUCATION:
A BRIEF HISTORY.................................... *Richard M. Abrams* 15

CAN UNIVERSITIES COOPERATE WITH
THE DEFENSE ESTABLISHMENT? .... *Carl Kaysen* 29

CONSEQUENTIAL CONTROVERSIES...................... *David A. Wilson* 40

THE PROGRAM

THE PHYSICAL SCIENCES AND MATHEMATICS .... *Edward Gerjuoy
and Elizabeth Urey Baranger* 58

ELECTRONICS AND COMPUTING.............................. *Leo Young* 82

DoD, SOCIAL SCIENCE, AND
INTERNATIONAL STUDIES .... *Richard D. Lambert* 94

SPECIAL PERSPECTIVES

THE MILITARY AND HIGHER EDUCATION
IN THE USSR .... *Julian Cooper* 108

THE VIEW OF THE BIG PERFORMERS....................... *Steven Muller* 120

THE GOOD AND THE BAD

THE GOOD OF IT AND ITS PROBLEMS .... *Richard D. DeLauer* 130

MILITARY FUNDING OF
UNIVERSITY RESEARCH .... *Vera Kistiakowsky* 141

BOOK DEPARTMENT 155

INDEX 201

BOOK DEPARTMENT CONTENTS

INTERNATIONAL RELATIONS AND POLITICS

JONES, ARCHER. *The Art of War in the Western World.* David G. Chandler.............. 155

MODELSKI, GEORGE and WILLIAM R. THOMPSON. *Seapower in Global Politics,*
1494-1993. Alan W. Cafruny 156

REISS, MITCHELL. *Without the Bomb: The Politics of Nuclear*
Nonproliferation. Michael Brenner 157

VIGOR, P. H. *The Soviet View of Disarmament;*
BLACKER, COIT D. *Reluctant Warriors: The United States, the Soviet Union,*
and Arms Control. A. Reza Vahabzadeh 158

WOOLEY, WESLEY T. *Alternatives to Anarchy: American Supranationalism since*
World War II. Frank D. Cunningham 159

AFRICA, ASIA, AND LATIN AMERICA

BANISTER, JUDITH. *China's Changing Population.* Sanjay Subrahmanyam 160

BULMER-THOMAS, VICTOR. *The Political Economy of Central America*
since 1920. John C. Beyer 161

EVANS, PAUL M. *John Fairbank and the American Understanding of*
Modern China. Hilary Conroy....................... 162

LEWIS, BERNARD. *The Political Language of Islam;*
MUNSON, HENRY, Jr. *Islam and Revolution in the Middle East.* Hamid Dabashi....... 163

STIVERS, WILLIAM. *America's Confrontation with Revolutionary Change*
in the Middle East, 1948-1953;
KIRISCI, KEMAL. *The PLO and World Politics: A Study of the Mobilization of*
Support for the Palestinian Cause. Charles D. Smith................................ 165

EUROPE

ASCHER, ABRAHAM. *The Revolution of 1905: Russia in Disarray.* Deborah Hardy 166

COHEN, MITCHELL. *Zion and State: Nation, Class and the Shaping of Modern*
Israel. Byron D. Cannon......................... 167

D'AGOSTINO, ANTHONY. *Soviet Succession Struggles: Kremlinology and*
the Russian Question from Lenin to Gorbachev. Jack L. Cross 168

ELLWOOD, SHEELAGH M. *Spanish Fascism in the Franco Era: Falange Española*
de las JONS, 1936-76;
POLLACK, BENNY with GRAHAM HUNTER. *The Paradox of Spanish Foreign*
Policy: Spain's International Relations from Franco to Democracy.
Philip B. Taylor, Jr........................ 169

KOLODZIEJ, EDWARD A. *Making and Marketing Arms: The French Experience and*
Its Implications for the International System. David B. H. Denoon 170

LEWIN, MOSHE. *The Gorbachev Phenomenon.* Scott Nichols 172

TUGENDHAT, CHRISTOPHER. *Making Sense of Europe;*
LEDEEN, MICHAEL A. *West European Communism and*
American Foreign Policy. Frederick J. Breit 173

UNITED STATES

DENHARDT, KATHRYN G. *The Ethics of Public Service: Resolving Moral Dilemmas
in Public Organizations;*
THOMPSON, DENNIS F. *Political Ethics and Public Office.* Janet K. Boles 174

GERTEIS, LOUIS S. *Mortality and Utility in American
Anti-slavery Reform.* Olive Checkland 174

JONES, CHARLES O. *The Trusteeship Presidency: Jimmy Carter
and the United States Congress.* Melvin Small 175

LANDAU, SAUL. *The Dangerous Doctrine: National Security and
U.S. Foreign Policy.* Cal Clark 176

LOWE, RICHARD G. and RANDOLPH B. CAMPBELL. *Planters and Plain Folk:
Agriculture in Antebellum Texas.* Rondo Cameron 177

PINDERHUGHES, DIANE M. *Race and Ethnicity in Chicago Politics: A Reexamination
of Pluralist Theory.* William M. Bridgeland 178

REINSCH, J. LEONARD. *Getting Elected: From Radio and Roosevelt to
Television and Reagan;*
TEIXEIRA, RUY A. *Why Americans Don't Vote: Turnout Decline in
the United States 1960-1984.* J. H. Bindley 179

ROBERTS, ROBERT N. *White House Ethics: The History of the Politics
of Conflict of Interest Regulation.* Fred Siegel 180

ROSSWURM, STEVEN. *Arms, Country, and Class: The Philadelphia Militia
and "Lower Sort" during the American Revolution, 1775-1783.*
Anne M. Ousterhout 181

SMITH, J. OWENS. *The Politics of Racial Inequality: A Systematic
Comparative Macro-Analysis from the Colonial Period to 1970;*
BUMILLER, KRISTIN. *The Civil Rights Society: The Social Construction
of Victims.* Howard Winant 181

SOCIOLOGY

BENNETT, GEORGETTE. *Crimewarps: The Future of Crime in America.*
Alvin Boskoff 183

DANIELS, NORMAN. *Am I My Parents' Keeper? An Essay on Justice
between the Young and the Old;*
RIVLIN, ALICE M., and JOSHUA M. WIENER with RAYMOND J. HANLEY
and DENISE A. SPENCE. *Caring for the Disabled Elderly:
Who Will Pay?* Francesco Cordasco 184

GIBSON, MARGARET A. *Accommodation without Assimilation: Sikh Immigrants in
an American High School.* Lelah Dushkin 186

GOYDER, JOHN. *The Silent Minority: Nonrespondents on
Sample Surveys.* George H. Conklin 187

JOYCE, ED. *Prime Times, Bad Times;*
BOYER, PETER J. *Who Killed CBS? The Undoing of America's
Number One News Network.* Fred Rotondaro 187

WUTHNOW, ROBERT. *The Restructuring of American Religion: Society and
Faith since World War II.* John A. Coleman 188

ECONOMICS

ASCH, PETER. *Consumer Safety Regulation: Putting a Price on Life and Limb.* M. O. Clement 190

BROAD, ROBIN. *Unequal Alliance: The World Bank, the International Monetary Fund, and the Philippines.* Colin D. Campbell 191

HEWETT, ED A. *Reforming the Soviet Economy: Equality versus Efficiency.* Holland Hunter 192

QUINN, DENNIS PATRICK, Jr. *Restructuring the Automobile Industry: A Study of Firms and States in Modern Capitalism.* Ashok Bhargava 192

TEMIN, PETER. *The Fall of the Bell System.* Bernard P. Herber 193

WINFIELD, RICHARD DIEN. *The Just Economy.* France H. Conroy 194

PREFACE

The military and universities: two great institutions of contemporary America; how are they related? In the past half century, they have both become vastly more important and powerful in the life of the United States. Because of the densely technical character of the American economy and life-style, higher education has become the sine qua non of ready entry and full participation. The requirements of science and technology include advanced education of scientists and engineers. So enrollments in universities have increased enormously since World War II. Universities, with their aggregates of talent and their control of credentials, occupy a crucial position in this nation.

The military equally has become a major element of the American scene. The military budget continues to consume a significant proportion of the gross national product, providing employment to millions as soldiers and sailors, civilian officials and defense contractors. In the world since 1945, profoundly influenced by the United States, the strategic role of the American military is obvious. So it is also the case that the military occupies a crucial position in the life of contemporary America.

It is not a surprise that these two segments of our pluralistic land are interrelated in important ways. Fundamental in this interrelationship is the military's dependence upon elaborate and sophisticated technologies to structure its forces and enhance the relatively tight supply of manpower with vast and effective firepower. Thus scientific research, technical advance, and system development are central. A large part of the nation's scientists and research engineers are to be found in the labs, offices, and classrooms of universities, and the military services have consistently relied upon the university community to meet its needs. Universities also produce the people— engineers, scientists, and managers—who fill positions in the services, the Department of Defense (DoD), and the defense industry.

The military is in part a vast educational institution, training men and women in the broad range of skills and disciplines appropriate to the complex needs of these times. It includes an extensive higher-education element. The three service academies educate as undergraduates the heart of the professional officer corps. They also support a number of graduate-level institutions as well as specialized advanced training facilities throughout the country. On civilian campuses, the services sustain the voluntary Reserve Officers' Training Corps, which teaches undergraduates military science and produces career officers as well as officers for temporary duty and reserve strength. In addition, many men and women in military service are placed in universities to pursue advanced degrees. Educational support is an important benefit for service men and women and, of course, since World War II, for war veterans.

Clearly, the military values education and the academic establishment. Whatever the conflicts, this aspect of the university-military relationship persists. In some form, it is historically stable, finding its basis in the Land-Grant College Act of 1862. It is also linked to the other important aspect, scientific research. The combination of research and graduate education that is characteristic of American universities means that continued research "is essential to training scientists and engineers."[1] Since the defense effort draws upon 14 percent of U.S. scientists and engineers, sustaining the educational capacity is a central purpose in the research program.[2]

Defense research is a part of the DoD Research, Development, Testing and Evaluation (RDT&E) program that constitutes about 10 percent of the defense budget. Universities play a significant role in the program, but in magnitude it is not large. In the years from 1960 to 1986, funding obligations from RDT&E to universities have ranged from 4.5 to 2.4 percent of the total. In 1986, the obligations to universities were 3.2 percent of the total RDT&E budget. (Figure 1.)

RDT&E is conceptualized as a structured process whose foundation is the so-called technology base, comprising basic research and exploratory development, or applied research. Universities are principal performers in the tech base. In 1986, 22 percent of tech-base funds went to universities, increased from a low of 11.3 percent in 1973. (Figure 2.) In 1986, two-thirds of the money allocated to universities was in the tech-base category. Although some of the other one-third was certainly involved in basic research, universities also performed early development work.

From the point of view of universities, DoD programs are also significant but not overwhelming. Table 1 shows that DoD obligation of funds over the 16 years from 1970 to 1986 ranged from 8.4 to 16.4 percent of federal research and development funds to universities.

What is clear is that universities and the military have been involved with each other in increasingly complex ways for many decades. The relationship vis-à-vis both education and research has been productive and satisfying to some. At the same time, it has puzzled and disturbed others. Because of its duration, complexity, and, perhaps, incongruity, it is worthy of a closer look. This volume provides a number of articles that take that closer look from a variety of perspectives. Each of the authors brings a special set of qualifications for the task.

Professor Richard Abrams is a well-known historian of the United States whose recent investigations have been directed to questions of the place and effect of the military in American society.

Professor Carl Kaysen, an economist, has been a student of and participant in the making of U.S. military policy for decades. Recently he chaired a committee that reviewed the military presence at the Massachusetts Institute of Technology, an institution with large and complicated relationships to the world of defense.

Dr. Leo Young, an electrical engineer, was the 1980 president of the Institute of Electrical and Electronics Engineers and served as director for research and laboratory management in the Office of the Secretary of Defense from 1981 to 1986. He is author of an extensive list of research publications.

Professor Elizabeth Baranger, a physicist, has a long-standing interest in the policy aspects of physics, a discipline central to the military-university relationship.

Professor Edward Gerjuoy, a physicist as well but also a lawyer, has been a regular participant in the American Physical Society and the American Bar Association activities related to policy, law, and science.

1. U.S. Department of Defense, *Report on the University Role in Defense Research and Development,* prepared for U.S. Congress, Committees on Appropriations, Apr. 1987, p. 5.

2. Ibid., p. 3.

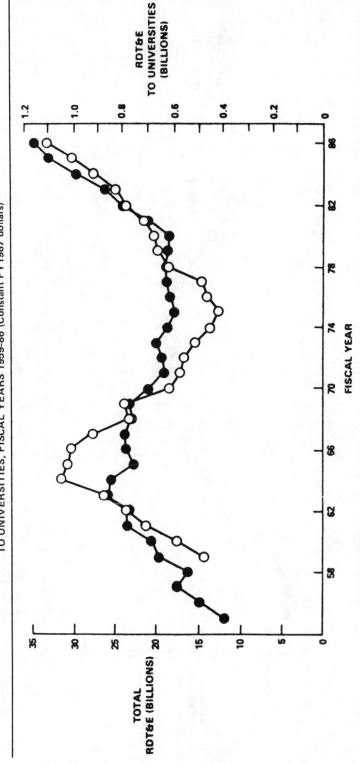

FIGURE 1

THE HISTORICAL TREND IN TOTAL RDT&E FUNDING, FISCAL YEARS 1955-86, AND IN RDT&E FUNDING TO UNIVERSITIES, FISCAL YEARS 1959-86 (Constant FY1987 dollars)

SOURCE: Department of Defense, *Report on the University Role*, p. 8.
NOTE: The dark points correspond to the left-hand vertical scale; the open points, to the right-hand vertical scale.

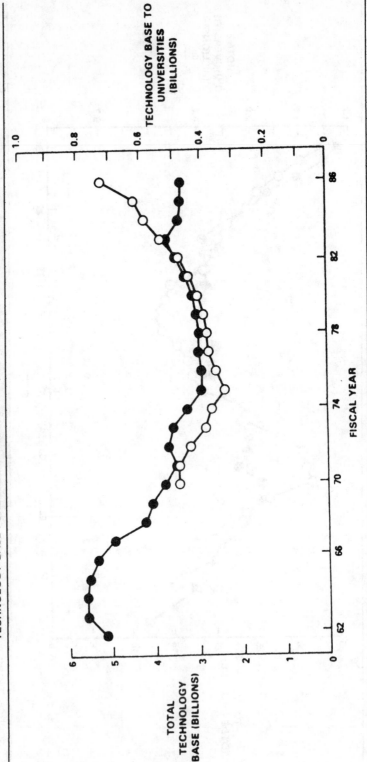

FIGURE 2

THE HISTORICAL TREND IN TOTAL TECHNOLOGY-BASE FUNDING, FISCAL YEARS 1962-86, AND IN TECHNOLOGY-BASE FUNDING TO UNIVERSITIES, FISCAL YEARS 1970-86 (Constant FY1987 dollars)

SOURCE: Department of Defense, *Report on the University Role*, p. 11.
NOTE: The dark points correspond to the left-hand vertical scale; the open points, to the right-hand scale. The technology base comprises DoD research and development budget categories 6.1 and 6.2.

TABLE 1
DoD OBLIGATIONS RELATIVE TO TOTAL FEDERAL OBLIGATIONS
FOR ACADEMIC INSTITUTIONS (Percentage)

Fiscal Year	Research and Development	Basic and Applied Research*	Basic Research
1970	14.6	13.5	15.7
1971	12.8	12.9	14.8
1972	11.4	10.8	12.7
1973	10.6	9.5	10.9
1974	8.9	8.5	9.3
1975	8.4	7.9	8.4
1976	9.4	8.5	8.3
1977	9.4	8.6	9.1
1978	11.3	8.3	9.5
1979	11.3	8.1	8.6
1980	11.6	8.5	9.0
1981	12.8	9.3	9.8
1982	14.4	10.2	11.2
1983	14.6	10.6	11.6
1984	14.9	10.7	11.5
1985	14.9	10.3	10.2
1986	16.4	12.0	12.3

SOURCE: U.S. National Science Foundation, Division of Science Resources Studies, *Federal Funds for Research and Development, Detailed Historical Tables: Fiscal Years 1955-1987*, Aug. 1986.

*As in National Science Foundation reports, DoD exploratory development (budget category 6.2) is regarded as applied research, and the sum of basic and applied research is the DoD technology base (budget categories 6.1 + 6.2). See U.S. Department of Defense, *Report on the University Role in Defense Research and Development*, prepared for U.S. Congress, Committees on Appropriations, Apr. 1987, p. 16.

Professor Richard Lambert, a sociologist, is a long-time leader in the social sciences, particularly those fields related to foreign area and language studies, subjects of importance to the place of the United States in the world. He has been effective in developing advice for DoD and other federal agencies on these matters.

Dr. Julian Cooper, lecturer in Soviet industry and technology at Birmingham University, brings a unique perspective as a specialist on the USSR.

Dr. Steven Muller, a political scientist, as president of The Johns Hopkins University—the largest academic performer of defense research in the nation—brings the perspective of a major university with substantial involvement with military organizations.

Dr. Richard DeLauer is of that important class in American society, scientist-industrialist. As a former executive of a high-tech defense contractor firm and also as the under secretary of defense of research and engineering in the first Reagan administration who generated and presided over significant growth of university-military cooperation in research, he brings a strongly affirmative point of view to these issues.

Professor Vera Kistiakowsky, a physicist, is an active and articulate critic of U.S. arms policy and the military-university programs.

I bring to the question the points of view of a political scientist and administrator in a major university system with experience in dealing with the military on a number of issues, including efforts from DoD to restrict the openness of academic science.

Running through these articles is a sense that the relationships being discussed are somehow paradoxical, that common sense suggests that things should be otherwise. This feeling comes out in two themes. One is that the essential character or purposes of a university and that of a military organization are contradictory, so that the soul of a university cooperating with the military is threatened. The second is that finding ways of sustaining an effective civilian influence on the military—for example, in the form of Reserve Officers' Training Corps officers and academic scientists in the picture—is a good, American thing to do.

It is certainly the case that the degree of cooperation between these two sets of institutions in the United States is unusually high, even though Dr. Cooper's article shows that it is not altogether unique. But this kind of institutional mélange is the way things are done in this country, and it has been since the founding of the federal republic. Our two themes fit well within this context, which is freighted with institutional tensions, jurisdictional disputes, and arguments about state's rights and civil liberties.

In the spectrum of opinion, career experience, and interest, our authors cover a broad middle area. As far as university-military relations are concerned, we do not include extreme enthusiasts or extreme opponents. We are neither purely academic nor completely from the world of affairs. We are not all social scientists, physical scientists, or engineers. We thus succeed in providing not only information and argument but perspective as well. Although the articles necessarily overlap in the material considered, each is a unique expression. Collectively, they provide substantial illumination of the subject matter.

DAVID A. WILSON

ANNALS, *AAPSS,* **502,** March 1989

The U.S. Military and Higher Education: A Brief History

By RICHARD M. ABRAMS

ABSTRACT: From almost unnoticed beginnings in the Morrill Land-Grant College Act of 1862, the military-university relationship has, since 1940, become a major feature of American society. Putting higher education to the service of public priorities has longer and stronger roots in the American tradition than does the ideal of the university as a sanctuary for independent, critical scholarship and disinterested pursuit of learning for its own sake. Ironically, the university gained its principal claim to eminence in the American mainstream only in the early twentieth century when much of the nation's elite came to respect the ideal of autonomous, disinterested research and teaching within an academic sanctuary. Although the ideal continued to be honored as worthy, its approximation to reality faded egregiously after 1940. Its very importance for the achievement of public priorities, most conspicuously for national defense, led the university to accept inducements and constraints that pulled it notably away from its briefly assumed mission as a protected refuge for the dispassionate and critical study of science and society.

Richard M. Abrams is professor of history at Berkeley. He is the author of The Issue of Federal Regulation in the Progressive Era *(1963);* Conservatism in a Progressive Era *(1964);* The Burdens of Progress *(1978); and articles in journals including* The American Historical Review; Business History Review; The American Bar Foundation Research Journal; Stanford Law Review; *and the* Political Science Quarterly. *He is currently engaged in research on the militarization of American society since 1949. He acknowledges the helpful suggestions of John Heilbron, David A. Wilson, and Daniel Kevles, and the Institutes of International Studies and Governmental Studies at Berkeley for supporting grants.*

THE American military's relationship with universities and colleges began in almost an absence of mind. It arose from an afterthought stipulation in the Morrill Land-Grant College Act of 1862 that institutions to be financed under the terms of the act through income from the sale of federal lands must offer military training as part of the curriculum. For three-quarters of a century, little of note developed from this beginning, but following World War II, the university-military collaboration became a vital feature of American society. Amid the unending tensions of the postwar era, Americans' call for their universities to service national policy priorities—especially some that required or made use of secrecy and deception—would put at risk higher education's own priorities for promoting honest and independent scholarship and teaching.

The main purpose of the Land-Grant Act was to promote "agriculture and the mechanical arts," but because the Civil War was already eight months in progress, Congressman Justin Morrill of Vermont was able to persuade his colleagues to insert into the measure the brief clause "and including military tactics." "Military instruction," Morrill commented, alluding to the North's lack of able officers at the start of the war, "has been incorporated [in the act] . . . upon the attention of the loyal states [to] the history of the past year." As the young congressman saw it, moreover, a continuing source of ample military skills among the citizenry would surely be needed as the country grew, "to secure that wholesome respect which belongs to a people whose power is always equal to its pretensions."[1]

1. Quoted in James E. Pollard, *Military Training in the Land-Grant Colleges and Universities* (Columbus: Ohio State University Press, 1964), pp. 57-58.

That was sixty years before the Republican Party came to identify itself with minimal-state ideology. It was the Democrats who then stood for limited government, generally opposing such uses of national resources as the Morrill Act authorized. President James Buchanan, in fact, had vetoed a land-grant college bill, which had no military training clause in it, in 1857. But the more activist Republicans had come to national power in 1861 pretty much determined to use government to stimulate business enterprise, both agricultural and industrial. They turned easily to making public universities serve this purpose, as part of a broader agenda. In addition to speeding the land-grant college measure through the post-secession rump Congress, they enacted a protectionist tariff bill that also bore the name of the congressman from Vermont, a national banking bill, and a farmers' homestead bill to encourage rapid settlement of the remaining continental territories.

Although specific public service was something of a novel assignment for American colleges at the time, two important American traditions underlay the military training provision. The first was Americans' strong commitment to the citizen-soldier, aimed at minimizing the need for a standing army. Morrill was particularly concerned to avert the eventuality that the nation's growing military needs might lead to a large professional military establishment, hence his preference for giving young men soldiering experience in civilian-run colleges. The second was the nation's legendary pragmatism that judged the university as it did every institution, by its utility in the achievement of prevailing social purposes. When, eighty years after the Morrill Act, modern science would revolutionize military technology and, accordingly, military strategy and its relevance for American foreign relations, the second tradition in particular would tend to commit American higher education to the service of the state in unprecedented measures. How seriously that commitment has come to compromise the mission of the university as a center for dispassionate and critical study of science and society remains a troubling question.

The question is especially troubling because there is no solid tradition in America of the university as a sanctuary for dispassionate and critical study of science and society. The commonly lofted ideal of academia, as a refuge where the scholar's social responsibility consists principally in his or her independent choices of critical research and learned conclusions, dates in America primarily from a brief period in the early twentieth century. Reality and the ideal probably converged most closely only during the twenty years between the world wars, when the value of academic freedom was first forcefully articulated and came to prevail among important segments of the nation's elite. Whatever detached independence American academics may have enjoyed in the eighteenth and nineteenth centuries arose mainly from the limited perceptions of their usefulness, which confined them mainly to producing cultured gentlemen, ladies, and clergymen who would perpetuate prevailing manners and standards of excellence. The evidence suggests that the universities in the post-Morrill era rather welcomed the addition to such seemingly slight tasks of a responsibility to contribute to the nation's economic growth and military preparedness.

Later, when higher education first gained eminence in America as a vital social asset, that achievement owed specifically to the way academic faculty won recognition as scientists with a detached devotion to truth, aloof from the passionate preconceptions of everyday life. As the nineteenth century came to its inglorious end, the perceived failures of both the democratic process and the economic marketplace suggested that both government and business required a recondite expertise that lay beyond the ready reach of practical experience. The university beckoned. Acting on their new strength, faculty organized and professionalized, raising the German ideal of learning for its own sake as its model and treating it as a tradition. The founding of the American Association of University Professors in 1915 drew attention to college faculty as a distinctive professional identity at the heart of the university, successfully linking the principles of tenure and academic freedom with the essence of scholarly integrity while countering the peculiar American practice of treating college faculty as mere employees.

Crucial support came from the business community. It had become well understood that modern business required attention to fast-growing developments not only in science and technology but in business management. In the nineteenth century it was commonplace for businessmen to regard a college education as having small or even negative value for their male heirs. By the Progressive Era, courtesy of the corporation revolution and the ascendancy of consumer-goods industries, business professionalism had become fashionable, and college-level business schools offered the appropriate credentials. An elite sector of the business community therefore grew receptive to the efforts of college faculties to define for themselves a unique place in society, independent of church, state, and economic pressures, sheltered from the winds of power and fashion, free to probe the limits of knowledge. It was prepared to accept that scholarly inquiry most efficiently served society when unfettered by mandates to service particular public priorities. It paralleled the prevailing business view that the business corporation, conceived by the state originally for stipulated public purposes, most efficiently served society when free to maximize profits.

The military's ascension in American life is still more recent. Unlike the university, the military made its move only upon the perception that specific public priorities demanded its augmented services. That did not come until World War II and the perpetual preparation for war thereafter. Actually, what the scholars, engineers, and scientists had wrought in the way of weaponry had even greater impact than did merely the war and the threat of war. The advent of jet propulsion and nuclear power had at least as much to do with America's assumption of hegemonic responsibilities in Western Europe and the Pacific after the war as did deliberate policy choices impelled by liberal internationalist premises. Modern military technology revolutionized warfare, put long-term strategic considerations at the center of diplomacy, and transformed military personnel from servants of policy into indispensable experts in the shaping of policy.

In this transformation, military personnel and the civilian officers of government would become senior partners with university scholars. Within less than a generation after American higher education first established the ideal of the university as a sanctuary for independent and disinterested scholarship, urgent public priorities would impose on it compromising utilitarian demands. From the

small beginnings of the Morrill Act,[2] the university-military relationship would grow into a major phenomenon.

SMALL AND HALTING BEGINNINGS

For the first half century after the Morrill Act, the military-training obligation appears to have had little impact. The act left it to the states to determine how it was to be imposed. Most land-grant states made it compulsory for able-bodied men in their first two college years, although War Department neglect led some to give it up for many years. Curricula included course work on tactics, but most of the training in the nineteenth century seems to have amounted to a few hours of drill each week. The act, in any event, did not appear to aim at producing officers but only at the presumed benefits of exposing young men to the experience of issuing commands and practicing obedience. As one Army spokesman put it, there was a general "hope that possibly a body of educated men having some knowledge of the elements of military science may be available in time of war to furnish the subaltern officers of a volunteer army."[3]

The country's entry into imperial competition at the turn of the century focused attention on various proposals for organizing a substantial reserve and restructuring the relationship between the regular army and the state militias. A General Staff report in 1915 strongly favored

2. Most histories of education entirely neglect the military provision of the Morrill Act. Frederick Rudolph's excellent survey history, *The American College and University* (New York: Vintage, 1962), p. 252, omits, without even an ellipsis mark, the clause "and including military tactics" even when quoting the key paragraph of the act.

3. Pollard, *Land-Grant Colleges*, p. 63.

building a large reserve officer corps from an expanded military academy system. But the country's powerful tradition against a strong professional military— and the anticipated expense—foredoomed that hope. The General Staff settled for its second choice, a provision incorporated in the National Defense Act of 1916 for a Reserve Officers' Training Corps (ROTC) in civilian colleges and universities, beginning with the land-grant colleges and the several military institutions like Norwich, the Citadel, and the Virginia Military Institute. Military drill would yield to military education as Congress gave the colleges the charge of preparing a large aggregate of reserve officers. American entry into the World War, however, stifled the program before it began to draw breath.

The National Defense Act of 1920 put the program on more permanent footing, but by this time the legitimacy of requiring universities to service specific public priorities had come into question. The war had exposed raw fissures in what had been presumed to be an American consensus on public priorities and social values. Excepting, most notably, the administrators of the land-grant colleges, educators now were arguing that it was specifically the role of the university to stand apart from the society's principal institutions so that it might most freely explore and reveal all ethical and policy options. Wartime experience with repression of dissent and the Red Scare excesses during and after the war kindled a powerful movement to establish academic freedom as the keystone of university life. Arguments about how military training could contribute to character building, widely persuasive among educators in the nineteenth century when words like "character" and "good citizenship" met with

complaisant understanding, now suggested mean-spirited intent to coerce conformity. On the other side, the increasingly pluralist character of the culture inspired new efforts by self-appointed custodians of the fading conventions to reimpose the fading conventions on everyone. The war-revived military establishment counted itself among the custodians and thereby confronted the strong insurgent attitudes in the universities.

Unhappily for the military, the Age of Normalcy lacked noteworthy enthusiasm for public causes of any kind, whether of the variety that inspired red raids or alcoholic abstinence, corporation controls or militant Americanism. The Army had hoped to gain a long-denied legitimacy and access to the mainstream of American life by exploiting its momentary popularity as victor in the Great War, and as a model for true Americanism in that era of patriotic ambivalence. Authorization by the Defense Act of 1920 to establish ROTC units in any college or high school, and to run youth camps as well, had promised a measure of that access. But the business of Americans was business, the military as yet had little to offer to business, and business did not want to pay for expensive military training units in or out of the colleges. It did not help matters when military spokesmen, grossly miscalculating the public temper, told Congress how ROTC instructors would "bring to bear at numerous points of contact, the ethical influence of Army traditions and ideals"[4] and proceeded to denounce as "bolshevists" the numerous college faculty members who questioned

4. Quoted in Eugene M. Lyons and John W. Masland, *Education and Military Leadership: A Study of the R.O.T.C.* (Princeton, NJ: Princeton University Press, 1959), p. 45.

the appropriateness of military training on campuses.

Nor did the War Department help by claiming inaccurately that the Morrill Act required compulsory training for all able-bodied males in participating institutions. Wisconsin reacted to this aggressiveness by rescinding the requirement in 1923. Two other states followed a few years later. The Interior Department, which ran the Office of Education in those days, declared that merely offering military training satisfied the law, and in 1930 the Justice Department endorsed Interior's position. Not willing to yield altogether to so-called pacifists and other designated radicals, most state governments maintained compulsory military training in the land-grant colleges, and most organized educational associations continued to champion ROTC, annually extolling the virtues of the program as a preferred alternative to universal military training in service of the citizen-soldier tradition, while complaining annually that the War Department gave them scant support. In 1934, the U.S. Supreme Court in *Hamilton* v. *California* upheld the constitutionality of compulsory ROTC, setting the stage for numerous anti-ROTC campus protests, harbingers of the 1960s, that continued until the program was suspended by the Army in 1940. That was when Congress authorized the draft and the establishment of Officers' Candidacy Schools for regular-army enlisted men. By then, a more serious collaboration of the universities with the military establishment was about to begin.

World War II was a popular war, perhaps especially in the academic community, where liberal internationalist impulses had been strongest for entry against the fascist powers. There seemed little dissent during this war that the proper functions of a university included

making positive contributions to the military effort. President Roosevelt's executive order in June 1941 creating the Office of Scientific Research and Development endorsed, essentially for the first time, the view that the proper functions of government included support of basic research by university scientists. Although that office's mandate, of course, focused on national security priorities, its acknowledgment of the importance of pure science lent assurance to academic researchers that their work need not be confined specifically to military objectives.

From the onset of the war, the universities' science and engineering facilities took on much of the research that produced missile technology, gun sights, bomb sights, radar, the proximity fuze, and, of course, the atom bomb. More than 25 universities secretly took contracts to develop chemical and biological weapons.[5] The degree to which university facilities were used for military research and development set the United States apart from other industrial nations, where most such work was done in government or private installations.[6] It also departed sharply from the American practice during World War I, when no one appeared to know how to direct government funds into the private or academic sector, and, instead, individual civilian scientists and engineers were given military commissions and absorbed into the services.[7] Since the initiative for most of the new weapons during World War II came in

5. See, for example, Barton J. Bernstein, "The Birth of the U.S. Biological-Warfare Program," *Scientific American,* June 1987, pp. 116-21.

6. Cf. Charles V. Kidd, *American Universities and Federal Research* (Cambridge, MA: Harvard University Press, Belknap Press, 1959), p. 26.

7. A. Hunter Dupree, *Science in the Federal Government* (New York: Harper Torchbook, 1964), pp. 313-25. Dupree writes, "If science was to take a new place in the conduct of war, it had to do it in the

fact from university scientists and engineers, it seemed a natural step for the government to contract with the universities to administer the projects while providing vastly enlarged facilities on, near, or sometimes at a considerable distance from campus.

DEFINING THE
APPROPRIATE RELATIONSHIP

From the war's beginning, it was generally assumed that the close, mutually supportive relationships between the military and the state, the university and the state, and the military and the university would continue after the war. Among other things, the assumption was built into the Office of Scientific Research and Development. As the war neared its end, Edward L. Bowles, science adviser to Secretary of War Henry Stimson, called for "an effective peacetime integration" of the military with the resources of higher education. "Not only is there a great opportunity to underwrite research for its direct contribution to the nation's welfare," he wrote, ". . . but the opportunity exists to encourage the training of brilliant minds and to instill in them a consciousness of their responsibility to the nation's security."[8] The onset of the

---

field of weapons research, and the armed services were jealous guardians of their own preserves." Ibid., p. 313. World War I did inspire the founding of the National Research Council, which brought together government and military officers, business executives, and scientists from private and academic institutions. But although its director, physicist Robert A. Millikan of the University of Chicago, sometimes recruited university personnel for particular projects—for example, sonar—with no important exceptions universities themselves neither took on war contracts nor administered military research undertaken by individual faculty members.

8. Quoted in Clayton R. Koppes, *JPL and the American Space Program: A History of the Jet Propulsion Laboratory* (New Haven, CT: Yale University Press, 1982), p. 26.

Cold War assured that government leaders would stress higher education's obligation to service national security needs. The heavy federal funding needed for modern scientific research more or less guaranteed that university faculty would be receptive.

All the same, after the war, when the consensus on foreign policy eroded and conscientious anxiety over their role in the nuclear arms race grew within the science community, some grumbling arose among scholars that their function as educators stood to be corrupted by continuing to accept funding for military assignments. While never denying their responsibilities to the public interest, as defined by public policy, many universities moved to protect the academic community from inappropriate entanglement with the state. Just what was inappropriate was never completely established, but research initiatives funded without peer review offered one criterion. Secrecy provided another. In 1946, Harvard president James Conant, himself an important science adviser to the government during the war, promulgated rules against university sponsorship of classified or secret research. That policy caught on elsewhere. But other universities—notably the Massachusetts Institute of Technology (MIT), just down the road from Harvard—for many years continued to permit classified research and even classified doctoral theses. MIT president James R. Killian, Jr., expressed great discomfort over such work but argued that the urgencies created by the Cold War made it necessary. "We have recognized," he said in 1953, "an inescapable responsibility in this time of crisis to undertake research in support of our national security which under normal conditions we would choose not to undertake. . . . When these conditions no longer hold, we shall withdraw from classified emergency research with enthusiasm and relief."[9] "Normal conditions," however, never returned.

Killian, later the first science adviser to the president, under Eisenhower, spoke of the 1940s and early 1950s as "a golden era in government-university cooperation."[10] The cooperation required important compromises. Apart from the secrecy, which was at odds with the character of a model academic environment, some of the military work took university scientists and engineers rather far into activities more suitable for industrial plants. Cal Tech's Jet Propulsion Laboratory (JPL), for example, did the basic and developmental research, then often produced the equipment and trained military personnel to use it. But most people seemed satisfied that the work was usually done at such off-campus facilities. As Killian commented:

The academic institutions responded by inventing novel ways to serve government without distorting their prime functions as educational institutions. The summer study projects, the special research centers, and the large off-campus interdisciplinary laboratories such as Lincoln, JPL, and the Applied Physics Laboratory, all managed by universities, are examples.[11]

By being willing to take on large military contracts in times of need, Killian remarked, the universities "made possible the maintenance of a scientific 'fleet in being' of great importance to national security."[12]

Using the model of the university as a detached community of scholars whose dedication to the honest pursuit of truth mandates complete openness, the argument against permitting secret research under university auspices had considerable appeal. On the other hand, perceived national needs and the American pragmatic tradition also had power. As one student of the problem wrote in 1959:

A few universities, among them Harvard, assess the deleterious effect of secret research as being so significant that they refuse to accept any classified projects. If this practice had been followed by all universities during the postwar years, either the national security would have been endangered by their failure to have the research done or competent faculty members would have been taken from universities to conduct research elsewhere.[13]

On the whole, a sense of "responsibility to the nation's security," together with the fact that modern research required funding in a magnitude not to be forthcoming from educational institutions or even from private industry, helped motivate most to continue essentially the wartime relationship. As Martin Trow has written, "The generous [federal] funding of scientific research after World War II played a major role in the explosion of knowledge in the United States and the rise to preeminence of American scientific disciplines in the world community."[14] During especially the first two postwar decades, the university-military connection underwrote those achievements. The work at JPL, at the Berkeley, Livermore, and Los Alamos laboratories of the University of California, and at the Instrumentation Laboratories at MIT, among others, had accustomed engineers and scientists to the new scale of equipment that postwar science required. The broad scope given

9. Quoted in Jack H. Nunn, "MIT: A University's Contributions to National Defense," *Military Affairs*, 43:124 (Oct. 1979).

10. Quoted in Koppes, *JPL*, pp. x-xi.

11. Ibid.

12. Nunn, "MIT," p. 124.

13. Kidd, *American Universities*, pp. 37-38.

14. "The Public and Private Lives of Higher Education," *Daedalus*, 2:113 (Winter 1975).

in Department of Defense (DoD) grants for basic research enabled university researchers to minimize the specifically military significance of their work. Mechanisms to guarantee peer review of contract awards added legitimacy to research broadly chartered by the nation's military-oriented agencies. They lent assurance to scholars that the federal grants compromised neither the quality of the scientific work supported nor their independent choices of inquiry. Projects for which the government required secrecy came increasingly to be confined to off-campus laboratories federally owned and financed but administered by universities, adding to the sense that military-oriented research left normal campus functions intact. By the 1960s, there were at least twenty Federally Funded Research and Development Center (FFRDC) laboratories sponsored by the DoD, the Atomic Energy Commission, or the National Aeronautics and Space Administration (NASA) and operated by 16 individual universities and several university consortia. An additional twenty or so other university research centers drew all or nearly all their financing from federal sources, with the DoD, the Atomic Energy Commission, and NASA accounting for most of it.[15] As of 1968, the DoD listed 92 universities among the top 500 prime contractors for research, development, testing, and evaluation.

15. Michael Klare, comp., *The University-Military Complex: A Directory and Related Documents* (New York: North American Congress on Latin America, 1969), pp. 52-54; Dorothy Nelkin, *The University and Military Research: Moral Politics at M.I.T.* (Ithaca, NY: Cornell University Press, 1972), pp. 28-32; Dean C. Coddington and J. Gordon Milliken, "Future of Federal Contract Research Centers," *Harvard Business Review*, Mar.-Apr. 1970, pp. 104-5. Depending on how one defines a Federally Funded Research and Development Center, the figures could be a bit higher.

Advantages to the military were, of course, substantial. First, the war experience showed that the military needed access to all the scholars that the country had who were at the leading edge of research and who did their work in an unregimented environment where the humanities as well as the social and physical sciences thrived. Second, academic researchers cost less than did those in private industry, as did most academic research facilities, while, on the other hand, civil service constraints that would have applied to government-run facilities did not get in the way of personnel recruitment. Third, having research and development (R&D) done at universities and other nonprofit FFRDCs obviated, in theory, potential conflicts of interest when it came to hardware production. Finally, just as the universities feared they would lose many of their best faculty if they severed connections with military-oriented research, so the military feared it would not enjoy the services of many of the best minds if it chose to confine its contracts and funding to private or government facilities.

The last probably was most important. In 1953, the DoD stated as its principal reason for funding university R&D the need

to maintain effective contact between the Armed Services and the scientific fraternity of the country, so that the scientists can be legitimately encouraged to be interested in fields which are of potential importance . . . to national defense, so that the entire scientific strength of the country could be brought to bear promptly and effectively in case of a severe emergency, so that the Services are continuously and growingly aware of scientific developments and of the value to them of scientific activity, and so that the scientists and the research administrators can contribute an important element of intellectual leadership within the Armed Services.[16]

Thirty years later, as part of the Reagan administration's dramatic military build-up, DoD used the same rationale in promoting its University Research Initiative. That program included the funding of new equipment at universities and the funding of sabbaticals for faculty to work in defense labs and later to continue the research back in their university facilities.[17]

SHIFTS IN FUNDING

Although the 1980s has witnessed a tightening of the military-university relationship, on the whole military-related funding has declined as a proportion of all federal funding of higher education. In the first postwar decade, about 86 percent of all federally funded R&D, and more than 90 percent of federal R&D support for the physical sciences in the universities, came from DoD and the Atomic Energy Commission.[18] The high figures, of course, reflected Congress's unwillingness at the time to provide federal aid to higher education except for defense purposes. Eventually, other federal funding agencies would greatly reduce DoD's proportionate role in the universities, while making

16. U.S. National Science Foundation, *Government-University Relationships in Federally Sponsored Scientific Research and Development* (Washington, DC: Government Printing Office, Apr. 1958), p. 10.

17. See, for example, "Pentagon Seeks to Build Bridges to Academe," *Science,* 19 Apr. 1985, p. 303; "Star Wars Grants Attract Universities," ibid., p. 304; "Enhancing Basic Research," *Aviation Week & Space Technology,* 6 May 1985, p. 11; "DOD Program Proves Attractive," *Science,* 4 June 1985, p. 129.

18. U.S. National Science Foundation, *Federal Funds for Research and Development for FY1978, 1979, and 1980* (Washington, DC: Government Printing Office, 1979), p. 6.

universities the prime recipients of federal support for basic R&D.[19] In 1950, Congress established the National Science Foundation; in 1958, after Sputnik shook the country, congressional obstruction of federal aid to higher education softened, leading to passage of the National Defense Education Act to underwrite foreign language study as well as engineering and the sciences; in 1965, creation of the National Endowments for the Arts and for the Humanities moved federal aid further from the military center, as did the Education Amendment Act of 1972. To date, the federal government remains the largest single funding source for academic R&D. On the average, it has supplied two-thirds of all academic R&D funding since 1960, peaking in 1965 at 73 percent, with lows of 63 percent in 1960 and 64 percent in 1985. The universities' share of all federal outlays for basic and applied R&D about doubled in the two decades after 1960, rising from 16 to 31 percent, while the share of all federal funding granted to FFRDCs declined slightly.[20] At the same time, DoD's share of all federal funds for basic research at universities dropped from about 25 percent in 1960 to a low of 8 percent in 1976 and about 12 percent in 1985, while the National Institutes of Health's share has ranged from 33.4 percent in 1967 to almost half by 1985.[21]

19. U.S. National Science Foundation, *Science and Technology: Annual Report to the Congress* (Washington, DC: Government Printing Office, June 1980), esp. pp. 22, 24.

20. National Science Foundation, *Science and Technology: Annual Report to the Congress* (Washington, DC: Government Printing Office, Aug. 1978), pp. 46-47.

21. Calculated from U.S. National Science Foundation, *Science Indicators: The 1985 Report* (Washington, DC: Government Printing Office, 1985), app. tabs. 2-5, 5-20.

Americans' increased willingness to use federal resources to underwrite research in all fields, especially in the biological and behavioral sciences, helps to explain the shift. Since the early 1960s, the National Institutes of Health has been the largest single sponsor of R&D at the universities; the National Science Foundation has been second. Yet it must be understood that the kind of research done under the auspices of the National Science Foundation and the National Institutes of Health does not always differ substantially from that done under DoD sponsorship. The charge of the three federal agencies, as with that of NASA and the Department of Energy, is the same: to promote work of potential benefit to the national security, a concept that has come to include leadership not merely in weaponry but also in industrial innovation. By the 1980s, the State Department was as likely as DoD to insist on restricting accessibility of university research on grounds that release of new technology information could jeopardize the nation's economic strength, thereby undermining national security.

THE DIMMING OF DISTINCTIONS

The physical and natural sciences have by no means been the only areas of DoD funding at the universities. Given the nature of modern military technology, the distinction between what belongs to the military and what to the civilian sector has dimmed. Since World War II, civilians and military officers of necessity have continually crossed over the lines in both management and policymaking responsibilities. It came to be assumed that in addition to the traditional mastery of combat skills, the new professional military officer would participate in defining

"the nature of the nation's security tasks, especially their politico-military dimension; . . . applying science and technological knowledge to military matters; . . . [and] advising foreign military establishments."[22] That gave academia additional tasks. In 1950, General Dwight Eisenhower wrote in a memorandum for Defense Secretary James Forrestal that "under present conditions," all regular officers had to have "a background of general knowledge similar to that possessed by the graduates of our leading universities."[23] For this reason and others, the services turned increasingly to ROTC and to postgraduate education for its officers at the civilian universities, rather than to the military academies and postgraduate institutions like the National War College. Obversely, since the 1940s, DoD has sponsored summer studies programs designed to bring academic scholars together with the military at home and abroad, both to acquaint military officers with advanced thinking in organization and technology and to instruct the academicians about military needs. Since 1950, DoD—along with private foundations such as Rockefeller and Ford—has also underwritten dozens of national security studies programs at the universities, as well as at think tanks closely tied to the individual services and DoD. These think tanks include, for example, Rand, which has ties with the Air Force; the Research Analysis Corporation, which has a close relationship to the Army; the Operations Evaluation

22. Amos A. Jordan and William J. Taylor, Jr., "The Military Man in Academia," *The Annals* of the American Academy of Political and Social Science, 406:130 (Mar. 1973).

23. Quoted in John W. Masland and Laurence I. Radway, *Soldiers and Scholars: Military Education and National Policy* (Princeton, NJ: Princeton University Press, 1957), p. 28.

Group, associated with the Navy; and the Institute for Defense Analyses, associated with the DoD.[24]

During the 1960s, when consensus on foreign policy collapsed, critics of the American military role in Vietnam focused intensely on challenging the university-military connection. DoD sponsorship of university R&D and the presence of ROTC units on campus took the brunt of the criticism. As criticism evolved into violence in the years 1967-70, fires and explosions damaged or destroyed some 200 ROTC buildings on campuses across the nation.[25] Bowing to what administrators took to be the prevailing views of faculty and students, several universities—for example, Stanford and Harvard—ended their officer-training programs, more than eighty made them voluntary, and for a number of years enrollment dropped more than 50 percent.[26] Other campuses, such as the University of California, Berkeley, rescinded course credit for ROTC curricula, although Berkeley a few years later restored credit under new arrangements whereby DoD agreed to permit military teaching appointments and course syllabi to undergo regular faculty scrutiny. A few universities, also bowing to antimilitary pressure on cam-

pus, severed connections with some of their FFRDC divisions, such as MIT's Instrumentation Lab and the Stanford Research Institute. Unanswered criticism of inadequate oversight by the universities of the FFRDCs they managed, including failure to enforce principles of academic freedom and intellectual honesty, left them vulnerable.

None of this, however, seriously impeded the momentum of growing state uses of the university for public purposes. While the state employed the university to produce basic and applied research, technology, political analysis, security studies, and other services on behalf of national defense priorities, it also made use of the university to help solve the nation's racial and ethnic problems. The university thereby became a target for attack from sources across the political spectrum.

Those on the left condemned the services provided for national security as skewing research and teaching away from the scholarly objectivity that defined academic integrity. Those on the right noted how federal affirmative action hiring and admission requirements designed to redress historic patterns of racial oppression undermined standards of excellence that also defined academic integrity. Attacking at the center were those who argued that scholars can never remain detached from social priorities, thereby directly challenging the very ideal of the university as a sanctuary for independent inquiry. On one side, the head of NASA during the 1960s, James Webb, insisted that the universities had an obligation to support the state and the capitalist system, and he sought to make grants to faculty contingent on "an increased awareness... of their societal responsibilities in the attainment of national goals."[27] A critic

24. See Gene M. Lyons and Louis Morton, *Schools for Strategy: Education and Research in National Security Affairs* (New York: Praeger, 1965); Masland and Radway, *Soldiers and Scholars;* Lyons and Masland, *Education and Military Leadership.*

25. Kirkpatrick Sale, *SDS* (New York: Vintage, 1973), pp. 406, 427, 503, 632, 724.

26. Sale says 30 units were dropped between 1966 and 1970. Ibid., p. 9. Between 1966 and 1972, despite the ROTC Vitalization Act of 1964, which enabled the Army and Air Force to match the Navy's already attractive program of full-tuition scholarships, enrollment in the Army ROTC dropped from about 177,000 to about 50,000. Jordan and Taylor, "Military Man in Academia," p. 135.

27. Quoted in Koppes, *JPL,* p. 144.

on the other side, urging scientists to take a "moral" stand against nuclear weapons research on campus, insisted in 1964, "The science expert . . . has no real hope of keeping out of politics."[28] The observation may well fall in the category of self-fulfilling admonitions.

Demands that scholars make judgments about the social and moral implications of their work, whether conforming or contrasting with prevailing national goals, carried with them hints of limiting inquiry to acceptable subjects and, perhaps, acceptable conclusions. To pick and choose among government-funded projects on the basis of what was or was not compatible with conscience or social-policy preferences implied the division of scholarship along political lines. And yet, beginning especially in the 1960s, many scholars found themselves impelled to make just such choices. Those whose political judgment led them to believe that enhancement of military capabilities encourages militarization of policy pressed for dissociation of the university from so-called war work. At the risk of impinging on academic freedom, they were prepared to place scholarship under the test of severe public scrutiny and constraint. For example, on the strength of the antiwar agitation that stirred campuses everywhere in the late 1960s, faculty and student activists led MIT to set up a screening committee for all externally funded proposals, making one criterion for approval the degree to which a proposal enjoyed "favorable attitudes" within the campus community.[29] Such critics, of course, met resistance from (1) those who

believed that disengagement from ties with DoD would serve only to reduce academic input into defense strategies, (2) those who saw disengagement as impoverishing scientific research at universities, (3) those who saw clear threats to academic freedom, and (4) those with contrary political judgments about the value of military power for deterring war. Many of the latter saw no great reason not to volunteer their talents for classified research, covert intelligence work, or even, according to the U.S. Senate Church Committee findings in 1976, disinformation activities that entailed deliberate fabrication of reports for particular public-opinion effects.

Amid the student and faculty protests in the late 1960s that led a number of universities to sever their connections with FFRDCs, the *Air Force & Space Digest* commented: "If the universities turn their backs on the real world of international conflict, unpleasant as that world is, they will lose a major portion of the relevance they are so consciously seeking these days."[30] That was a fair-enough touch. But "relevance" may prove to be the bane of academic integrity. One of the ironic outcomes of the Vietnam-era agitation was the so-called Mansfield Amendment to the DoD authorization bill for fiscal year 1970, which precluded DoD funding of any research project unless it had "a direct and apparent relationship to a specific military function." Although intended to reduce DoD's role in the shaping of basic research, it is hard to imagine a more effective way of assuring that DoD's enormous financial resources would increasingly skew research toward defense priorities. Although Congress later modified the Mansfield constraint to permit funding of research

28. Warner Schilling, quoted in Nelkin, *University and Military Research*, p. 10.

29. Ibid., p. 91. See also Center for Strategic and International Studies, *U.S. Military R&D Management*, special report no. 14 (Washington, DC: Georgetown University, 1973), pp. 49 ff.

30. Nelkin, *University and Military Research*, p. 84.

that had a "potential relationship to a military function," throughout the 1970s and early 1980s not only did DoD guidelines for grant applications insist on "the relevance of the proposed research to the DoD mission," but the department also encouraged "preproposals" so that university researchers could modify projects when their initial ideas did not precisely meet DoD interests.[31] Worse, DoD explicitly rejected peer review of research proposals, substituting merit review, which, to curry favor with a broader congressional constituency, included such criteria as a balanced geographic and institutional distribution of awards.

CONCLUSION

In the postwar crisis that has never ended, the model of the university as an island of disinterested scholarship has taken a beating. The university's increasing commitment to public service since World War II not only has impelled scholars to perform tasks in secret, but, more important, has tended to bind them to uncritical postures on matters that they are known to be experts in, postures that in the normal course of academic life would border on intellectual dishonesty. Recent reports, essentially uncontradicted, that eminent scientists at the University of California's Lawrence Livermore Laboratories knowingly exaggerated the feasibility of President Reagan's Strategic Defense Initiative[32] exemplify the danger to academic integrity of the snug ties of research to public policy. In pursuit of federal funds after the hard times of the 1970s, universities throughout the country began actively to lobby Congress for research contracts even when there was strong likelihood that DoD would enforce secrecy and constricted access of scholars to the research and the findings.[33]

Altogether, it may be said that academic engagement with the defense establishment seriously compromised the critical independence of the university. In doing so, it may have undermined the university's most valuable quality—the capability of offering society disinterested criticism of prevailing outlooks and institutions—indeed, the very quality of the academic profession that first raised it to eminence in American life. On the other hand, the American habit of pressing the university into service on behalf of public policy priorities likely assured that outcome anyway. The responsiveness of biogenetic and social science researchers to popular views on sensitive social issues, such as race, gender relations, sexual preferences, and ethnicity, has shown similar tendentiousness. Knowledge is power; it would be surprising if the purse-holding public showed disinterest in the political implications of social and scientific research. That scientific inquiry and intellectual talent should follow the leads of money and politics is in any case hardly new, either in twentieth-century America or elsewhere and at other times in history. All the same, nothing has had the overall force of the defense establishment in redirecting basic and applied research, in putting limits on the free exchange of intelligence, in dampening discussion of the merits of research that has policy implications, or in converting scientists into policy advocates and scholars into entrepreneurs.

31. U.S. Department of Defense, *Report on the Merit Review Process for Competitive Selection of University Research Projects,* Apr. 1987.

32. See Deborah Blum, "UC, 'Star Wars' Data Hit as Summit Nears," *Bee* (Sacramento, CA), 4 Dec. 1987; idem, "'Star Wars' Lab Consigns Whistle-Blower to Limbo,"ibid., 6 Dec. 1987; idem, "X-ray Laser Unproven, Says H-bomb Creator," ibid., 10 Dec. 1987; "Briefing," *Chronicle* (San Francisco), 13 Apr. 88.

33. "Pork Barrel Science: No End in Sight," *Science,* 3 Apr. 1987, pp. 16-17.

ANNALS, *AAPSS,* **502,** March 1989

Can Universities Cooperate with the Defense Establishment?

By CARL KAYSEN

ABSTRACT: Deeply influenced by their experiences in World War II, the military services and the Department of Defense became important and supportive funders of academic science—chiefly the physical sciences—in the postwar period. They made an implicit contract with the academy, providing support under the going rules: the pursuit of new knowledge; investigator initiative; publication of the results; and some form of peer review as the allocative instrument. In return, the academy offered new science and trained scientists. The relation persists, but changes in the political context, the internal capabilities of the Defense Department, and the growth of other forms of support have changed it in a long-lasting way. Nonetheless, there are good reasons on both sides to continue the relation, and a willingness by both to accept the terms of the implicit contract would allow it to continue.

Carl Kaysen is professor of political economy in the Science, Technology and Society Program at the Massachusetts Institute of Technology. He has long been interested in the economics and politics of research.

IN the half century since the beginning of World War II, the answer to the question posed in the title has moved only a small way from "willingly and easily" to "not at all" on the spectrum of possible answers. But the net movement has clearly been in the latter direction, although neither steady nor smooth. This article examines the reasons for the change and, in their light, considers the likely future of the relation.

We focus on one aspect of the relation, defense as supporter of academic research, but two others are worth mentioning, if only to acknowledge them. First, the Department of Defense (DoD) and such military activities as are conducted in the Department of Energy (DoE), the National Aeronautics and Space Administration (NASA), and the Central Intelligence Agency—for short, the defense establishment—constitute a major user of trained scientists and engineers. The defense establishment and the major defense suppliers employ some 13 percent of all U.S. engineers and scientists,[1] and they have therefore a strong interest in the quality and quantity of the flow from the universities. In this, their relation to the academic suppliers is similar to that of other large users—the electronics industries, for example. There is one important difference: in the past, the perceived needs of the defense establishment have led to direct federal subsidies for the education of engineers and scientists; so far no other users have achieved a like result.[2] Second, through reserve officers' training programs, the universities recruit and supply a substantial proportion of the active-duty officer corps of all the services, as well as their reserves.[3]

It is appropriate to emphasize defense's role as research funder, both because research is one of the universities' central activities and because it is primarily in this role that the defense establishment connects with a substantial segment of the faculty, as distinct from the administrative apparatus, of the universities.

Like so much else that characterizes the current situation of American universities, especially their relations with the federal government, the role of the defense establishment in funding academic re-

---

1. In 1982, the Census Bureau, on behalf of the National Science Foundation (NSF), conducted a sample survey of those classified as "scientists, engineers and related occupations" in the 1980 census. In response to the choices offered regarding a question on what the major functional focus of their work was, 13 percent chose "defense." This information, which has not yet been published, was communicated to me by Dr. Carlos Kruytenbosch of the Science Indicators Unit of NSF. These are overall figures, covering the whole range of scientific and engineering disciplines. In particular industries and specialties, the proportions are much higher. For instance, the six most militarily dependent U.S. industries, employing 3.3 percent of the civilian labor force in 1982, employed about 21 percent of all engineers, 24 percent of all electrical engineers, 32 percent of mathematicians, and 34 percent of physicists. See Rebecca Blank and Emma Rothschild, "The Effect of United States Defense Spending on Employment and Output," *International Labor Review*, 124(6):693-94 (Nov.-Dec. 1985).

2. The National Defense Education Act was passed in 1958, in response to the Soviet's launching of *Sputnik* in October. Among the other provisions, it offered federal funding of loans to undergraduates, with emphasis on those studying science, engineering, and foreign languages, and fellowships for graduate students in the same areas, with provisions for loan forgiveness for those who went into teaching. See Ronald A. Wolk, *Alternative Methods of Federal Funding* (New York: McGraw-Hill, 1968).

3. In fiscal 1986, the contribution of service academies and the Reserve Officers' Training Corps to the officer corps of the three services was as follows: the Army received 1023 officers from the academies and 4669 from the Reserve Officers' Training Corps; the Air Force received, respectively, 938 and 2369; and the Navy and Marines received, respectively, 849 and 1364. Communication from the Public Information Office, First Naval District, Boston.

search has its origins in World War II. It can be traced specifically to the decisions made by the National Defense Research Committee to fund contracts for military research and even weapon development through, and initially in, universities. In part, the decision reflected the backgrounds of the committee's leaders— Vannevar Bush of the Carnegie Institution and previously the Massachusetts Institute of Technology (MIT), Karl Compton of MIT, James Conant of Harvard, and Richard Tolman of Cal Tech; in part, it was their considered judgment that it would be easier and quicker to mobilize scientific talent for war purposes by keeping people in place as much as possible, rather than trying to move them all to newly created laboratories in new locations. This device led both to funding small groups working in their usual academic locations—such as the initial group working on uranium at Columbia under Harold Urey and John Dunning or the expansion of E. O. Lawrence's work with the cyclotron at Berkeley—and to the creation of large new laboratories drawing on distant as well as local talent, such as the Radiation Laboratory at MIT to work on radar and the Applied Physics Laboratory at Johns Hopkins to develop proximity fuzes. The National Defense Research Committee's choice worked well, but it was not inevitable; a different choice might well have led to a radically different state of affairs today. Great Britain, for example, followed a substantially different path during the war, which has led to a much different situation with respect to university-defense relations today.

The emblem of academic science's and scientists' contribution to winning the war was the atomic bomb, especially in the public and congressional mind. But both the defense establishment and the academic community knew there was much more, from Atabrine and penicillin to naval fire-control systems and microwave communication. Further, they saw how close the other side had come, with pilotless planes, ballistic missiles, and jet aircraft engines, to outstripping us in vital areas of military technology.

By the end of World War II, their experience led the military services to view the community of academic science not only as a source of new ideas for weapons and even operational methods but also as a reserve of highly skilled and adaptive people, who could be mobilized for a wide variety of tasks should the need again arise. And the increasing differences between the United States and the USSR, which rapidly turned alliance into antagonism, soon made the need appear nearer rather than further in the future.

On their side, academic scientists enjoyed the stimulus and excitement of their war work, took justified satisfaction in their achievements, and savored the public acclaim received—novel as well as gratifying experiences for almost all of them. Academic administrators in general enjoyed the flow of funds. The wartime contracts had provided for institutional overhead as well as direct costs. Universities can always use more money, and the postwar continuation of the practice of providing overheads, as well as paying some part of faculty salaries and supporting graduate students as research assistants, was dazzlingly attractive.

After the war's end, the Navy continued to support research in the universities through the Office of Naval Research, established in 1946. Its support shifted from the design of specific weapons or the solution of problems with a direct military impact—understanding the transmission of sound in the oceans, for example—to the more general support of research in areas of physical science that seemed relevant to the Navy's potential concerns; and the criteria of relevance were very broadly interpreted. The other services followed the same path: the Air Force Office of Scientific Research was established in 1952; the Army's in 1958. The two later creations followed that of the National Science Foundation (NSF) in 1950, with a broad charter to support the advancement of knowledge; Alan Waterman, the NSF's first director, came to the task from the Office of Naval Research, where he had been the chief civilian scientific administrator. Later organizational rearrangements transferred funds and responsibility to agencies within the Office of the Secretary of Defense, such as the Advanced Research Projects Agency, founded in 1959, but the military services continued as major supporters of academic research.

The Atomic Energy Commission (AEC), a civilian agency, was created by Congress in 1946 to take over the assets and operations of the atomic bomb project from the Manhattan Engineer District of the Army: the Los Alamos laboratory, the production plants for fissionable materials at Hanford and Oak Ridge, and associated laboratories. It soon followed the Office of Naval Research's pattern of supporting basic research in universities, with a focus on nuclear physics and closely related subjects.

THE IMPLICIT CONTRACT

By the mid-1950s, a pattern of practices had developed that can be seen as an implicit contract between the defense establishment and the universities. In effect, the defense establishment undertook to support basic research in the universities—in practice, chiefly in mathematics and the physical sciences, with some small excursions into medicine and even smaller ones into social science—in a way that conformed to what can be termed the American academic style. The major elements that constituted the style were:

1. The purpose of research was to add to the corpus of scientific knowledge; academic publication was the typical outcome expected. This implied that no classified work would be done and that students, especially graduate students, would have access and opportunity to participate in ongoing research.

2. Primary initiative for research programs lay with individual scientists or small groups of scientists who made proposals to funding agencies.

3. Proposals had to conform to broad guidelines of acceptable scientific fields and administrative rules of duration, scale, and admissible items for expenditure set by funding agencies. Within these guidelines, the choice of proposals for funding was based primarily on scientific merit, though broad criteria of relevance to the purposes of the funding agency were sometimes involved, beyond the limitations of scientific field.

4. Judgments of merit were based on examinations by scientific peers of proposals for their intrinsic interest, promise, and the scientific capabilities of the proposers, evidenced by their past achievements. Scientific peer review was provided by some mixture of technically competent staff of the funding agencies and consultants drawn primarily from the world of academic science.

5. Funding in the form of contracts or grants went to the universities of the faculty or research staff on which the proposing scientist served, and it included overhead for the institution as well as the direct expenses of the project: salaries for

faculty and research assistants, often graduate students; salaries for support staff; equipment and supplies; travel.

On their side, the universities, in response, undertook first and chiefly to do what came naturally. This meant explicitly to encourage their faculty to participate in the research enterprise—or even demand that they do so by their criteria for appointments and promotions. The research faculty, implicitly and perhaps in large part unconsciously, also undertook to be there, furnishing the reserve capacity for much more directed research and even engineering efforts for military purposes, should the need arise. This commitment was both explicit and active, as well as existing as a reserve potential, for a small but important part of academic science, namely, the group who continued as consultants and advisers to ongoing defense activities with specific military goals. On a case-by-case basis that developed into a practice based on a pattern that started during the war, the universities also undertook to join with the defense establishment in creating a new species of research enterprise, the Federally Funded Research and Development Center (FFRDC). This was a freestanding, typically large laboratory, sometimes on or adjacent to the university campus, sometimes far removed, for which the university accepted managerial responsibility, and which the DoD and/or AEC funded. Such centers typically presented a different organizational model for funding research from the proposal-by-proposal researcher-initiated project. Supported essentially by block grants, the center itself, through its management structure, developed its research program. In practice, of course, the relative degree of top-down program management and bottom-up researcher-initiated proposals

has varied among centers and over time, depending on the center's mission, its management, and its relations with its sponsoring agency.

The first and still one of the largest of these was Los Alamos National Laboratory, for which the University of California has been the contractor from its inception in 1943.[4] Although its major function remains weapons development, Los Alamos has broadened its interests over time and now supports research in many fields of pure science, as well as applied work in areas other than nuclear weapons design. At a later stage, the NSF joined the defense establishment in sponsoring FFRDCs, some of these managed by university consortia, some by individual universities.[5]

Aside from filling the formal role of contracting agent and manager for the

4. See Leslie Groves, *Now It Can Be Told* (New York: Harper & Bros., 1962), p. 149, and fn. 6 of this article.

5. In 1987 there were 34 such centers: 17 administered by universities, 8 by other nonprofits, and 9 by industrial firms. The university-sponsored ones included 10 primarily funded by DoE, of which 5 were basic science research units for high-energy, nuclear, and plasma physics, 2 primarily for nuclear weapons research and design, and 3 others with a wider spectrum of activities in applied and basic research. Two were sponsored by the DoD, one by NASA, and the remaining four by NSF, all astronomical and atmospheric observatories. For the complete list, see *Federal Funds for Research and Development* (Washington, DC: National Science Foundation, 1987), 36:7-8. In fiscal 1985, the last year for which the data have been tabulated, all the FFRDCs received $5014 million and those administered by the universities—substantially but not exactly the same set as those referred to earlier—$2534 million; those by industrial firms, $1791 million; and those by other nonprofits, $689 million. The major supporters of university-sponsored labs were the following: DoE, $1848 million; DoD, $310 million; NASA, $231 million; and NSF, $85 million. In that year, Lawrence Livermore and Los Alamos National laboratories were the two largest of the university-administered centers, both

FFRDCs in return for a management fee, universities have varied in the extent and character of their involvement with the centers. In some cases, especially when the center was on or adjacent to the campus, university involvement has been deep. Center directors have been university faculty members; so have leading members of the scientific staff. Graduate students in the relevant fields typically have done their thesis research at the center and have been supported from its funds. The Stanford Linear Accelerator Center in Palo Alto and the Plasma Physics Laboratory in Princeton, managed by their respective universities, exemplify this situation. In other cases, with centers distant from university campuses, such as Los Alamos, no such relation obtains.

The FFRDCs sponsored by the defense establishment did not generally operate under the university rules described earlier. In particular, those with applied research and development tasks did classified work, and scientific publication was not their major measure of output. Thus some universities—though not all—tolerated classified work and limited access. Sometimes a distinction was maintained between on- and off-campus research, even when it involved a certain artificiality, and classified activities were allowed only off campus.

THE UPS AND DOWNS
OF MUTUAL RESPECT

The regime characterized previously was well established by the mid-1950s and functioned effectively for about a

administered by the University of California. Livermore received $631 million, of which $584 million was from DoE; and Los Alamos $538 million, $474 million from DoE. These data are drawn from unpublished tabulations provided by Dr. Kruytenbosch.

decade. It would be incorrect to say there were no strains before 1966, but, on the whole, they were minor. After the mid-1960s, the tone of relations changed greatly.

In 1949, when the call went out from the AEC to the academy for help in speeding up work on the hydrogen bomb, the response was mixed. Many important veterans of Los Alamos, such as Hans Bethe, refused, in part because of their individual doubts about the necessity of the program, in part because they were influenced by the negative recommendations on the desirability of a crash program by the AEC's General Advisory Committee, which included the prestigious scientific figures of Oppenheimer, its chairman, Fermi, Rabi, and Seaborg, as well as such leading scientists turned administrators as Conant of Harvard and DuBridge of Cal Tech. The defense establishment might well have seen this dissent as a failure of the academy to keep up its end of the bargain.

The rift was neither wide nor deep, though some of its consequences were lasting. More lasting was the turmoil caused in the academic world by the wave of spy catching, and hunting for the disloyal, the subversive, and the security risk. McCarthyism in all its manifestations—legislative investigative committees, loyalty oaths, an increasingly elaborate apparatus of security clearance—struck the academic community with particular force. For scientists, the climactic moment was the AEC's determination in 1954 that Robert Oppenheimer was a security risk and his exclusion from all advisory roles. While many individual scientists withdrew from classified work and governmental consulting in general, the broader relation between the defense establishment and academic science showed little change. Expenditures by the

DoD and AEC for the support of academic science continued to increase; the AEC remained the major supporter of nuclear and high-energy physics, still the most important and active part of the discipline and by far the most expensive, and the physicists within the universities gladly took the money. Despite the troubles, the implicit contract continued to function effectively.[6]

In the last twenty-odd years, the relation has been under much greater strain; downs of all kinds have outweighed ups. First, in both time and importance, was the escalation of the war in Vietnam and its increasing Americanization. Between 1963 and 1968, American military personnel in Vietnam increased from 30,000, nominally advisers to the South Vietnamese forces, to 500,000 troops frankly engaged in war with North Vietnam; there was also the involvement of additional Air Force and Navy personnel in the Pacific, producing sharp negative reactions at home. The war was unpopular in the country and particularly on American campuses, among both faculty and students who were liable to the draft. Beginning in 1966, teach-ins, strikes, and other demonstrations against the war appeared on campus after campus, and local manifestations of the defense establishment—Reserve Officers' Training Corps programs, adjacent FFRDCs, military-funded research in general—were frequently the targets of the demonstrators. The turmoil lasted for nearly six

years and transformed the previous relatively friendly and generally respectful relations between the academy and the defense establishment into one of substantial mutual hostility and contempt. It should be noted that both the targets and the sources of the negative attitudes on the government side were more the high officers of government, elected and appointed, than the professionals in the middle ranks of the military and civil services.

These events led to the discontinuance of Reserve Officers' Training Corps programs at some institutions, a greatly increased resistance to any kind of classified research in universities, and some instances of removal of on-campus FFRDCs to other sites or reorganization that severed their university connections.[7] While the level of generalized hostility to the defense establishment in university faculties grew, only a tiny minority expressed this hostility by refusing to accept funding from the defense establishment.

Another less direct consequence of the Vietnam war was the end of real growth in federal support for academic science. At the same time, the relative importance of the defense establishment as a source of support for academic research also

6. Between fiscal 1955 and 1966, total DoD obligation for research, development, testing, and evaluation doubled from $12.1 billion to $23.6 billion. The share going to academic institutions increased from 2.8 percent in 1960, the earliest year for which data are available, to 4.3 percent in 1966. See U.S. Department of Defense, *The University Role in Defense Research and Development*, prepared for U.S. Congress, Committees on Appropriations, Apr. 1982, tab. 1.

7. In 1972, Draper Laboratory, run for DoD and NASA by the Aero-Astro Department of MIT, was the object of demonstrations and a faculty resolution disapproving of the relation. As a consequence, the laboratory was reorganized as an independent not-for-profit corporation, and its formal connections with MIT severed. In Princeton, an Institute for Defense Analysis research center working on cryptographic problems that was located near the campus in space belonging to the university and leased to the institute was likewise the object of student demonstrations and faculty protests. It was relocated to a site several miles away. In 1970, Reserve Officers' Training Corps enrollments in U.S. colleges and universities were 123,000; by 1974, they had shrunk by more than half to 56,000; by 1985, they had recovered to 99,000.

declined, so that the strength and importance of the implicit contract was weakening. The National Institutes of Health
and the NSF outstripped the DoD and
AEC as sources of support for academic
science. The general inflationary and
fiscal pressures consequent on the war—
and President Johnson's refusal to increase taxes to pay for it—led to a
squeeze on all nonmilitary federal expenditures, and the growth in federal science
budgets fell far behind the growth in the
price level. Within the DoD, competing
demands for resources led to a decline in
total real expenditures for research, development, testing, and evaluation and an
even sharper decline in support for academic research.[8]

Fiscal pressure was the immediate
stimulus for these changes, but other
forces were probably at work. Public
faith in the unquestionable beneficence of
science was weakening, and public expressions of doubt were beginning to surface
within the scientific community. Congress

was no longer as ready as it had been
earlier to see each new bigger accelerator
as a good thing, and it began to ask
sharper questions about the usefulness of
high-energy physics.[9]

The government in Washington responded in its turn to the hostility on the
campuses. President Johnson expressed
his anger with the academic community
and even tried to cut specific appropriations for academic research. Campus
leaders appeared high on President
Nixon's notorious "enemies list." At least
one university head reported a threat to
withdraw all federal funding, a threat
that was never carried out.[10]

The Mansfield Amendment, attached
to the 1970 appropriation for the DoD,
required that the department support
research in the universities only if it was
of direct relevance to the military mission.
This further increased the hostile feelings
of academics toward the defense establishment. But the legislation reflected more
the sponsor's judgment that the services
were inappropriately involved in supporting pure science than hostility to the
universities; its chief sponsors, Senators
Mansfield and Fulbright, were themselves
former academics and strongly critical of
the U.S. role in the Vietnam war.

The Vietnam war wound down, but
the change in attitudes in both Washington and the universities persisted. President Nixon downgraded the status of his
scientific adviser and, early in his second
term, reorganized the office by assigning
its responsibilities to the then head of the
NSF and at the same time abolished the
President's Scientific Advisory Committee.[11] This committee, created by President Eisenhower as part of his response

8. In fiscal 1966, federal expenditures on research and development—in constant 1972 dollars—had reached $18.2 billion, compared with
$12.7 billion in fiscal 1960. They slipped over the
next nine years to $14.5 billion and had recovered
their previous peak only by 1983. See U.S. National
Science Foundation, *National Patterns of Science
and Technology Resources,* NSF 84-311, 1948.
Within the DoD, the peak year was fiscal 1967; the
trough was fiscal 1975, at 25 percent less. The
academic share was highest in 1965 at 4.5 percent; it
fell to 2.4 percent in 1974, recovered to a peak of 3.7
percent in 1980, and slipped back to 3.2 percent a
few years later. The absolute amount of DoD
support for academic research in constant dollars
fell by more than half between 1964 and 1975 and
recovered its 1964 level only in 1986. In 1967, DoD,
DoE, and NASA—the defense establishment—
spent 70 percent as much as NSF plus the National
Institutes of Health—the basic research establishment—on the support of academic research. In
1987, the ratio was 40 percent. See U.S. National
Science Board, *Science and Engineering Indicators,*
1987, app. tab. 4-11, p. 245.

9. See Daniel J. Kevles, *The Physicists* (New
York: Knopf, 1978), pp. 413-14.

10. Ibid., pp. 411-13.

11. Ibid., p. 413.

to Sputnik in 1957, symbolized the recognition of the need for the knowledge and independent judgment of the scientific community at the highest levels of government; its abolition was intended and received as a hostile gesture.

Here, again, forces deeper than the opposition to the Vietnam war in the academy, especially—but not only—in the elite institutions, were at work. Nixon's victory over Humphrey and his smashing defeat of McGovern represented a rightward shift in national politics, one that was intensified in the two victories of Ronald Reagan. While political opinion in the university world contains almost the whole American range, its distribution is far different. The academy—faculty and students—is substantially more liberal than the American voting public. The difference is even greater when the comparison is made between the political views of faculty members and those voters in general with broadly similar socioeconomic status and education.[12] Within the academy, natural scientists are about as liberal as faculties as a whole; social scientists and humanists top the scale in liberalism, while faculties in engineering, business schools, and schools of agriculture are at the bottom.[13] Thus the rightward shift in the national political spectrum deepened the division between Washington and the academy.

In recent years, a new set of specific irritations in the relations between academic science and the defense establishment has arisen. All reflect the renewal of the cold war at the end of the Carter administration and its sharp intensification during the first six years of Reagan's

presidency: the increased competition in strategic weapons, the escalation of the rhetoric of confrontation, the—mostly failed—efforts to exert economic pressure on the Soviet Union and its allies.

As part of its efforts to prevent the transfer of militarily useful technology to the Soviets, the administration sought in one case to classify ideas conceived by academic scientists not working under government contracts and even to require reporting of new discoveries in certain fields before publication. In another area, the DoD proposed to apply export controls on a scientific meeting, requiring the conveners to exclude scientists from certain foreign countries on the ground that their participation would constitute an unlicensed export of sensitive technology.[14] In neither case did the government succeed fully in carrying out its original intention, but the effect of breaking the rules with respect to free publication of new scientific knowledge remained.

Finally, there is the cluster of issues surrounding the Strategic Defense Initiative. At the broadest level, much of the relevant scientific and technical community in the academy has been amazed by the process that led to President Reagan's decision to make a commitment to the program and specify its goals without even drawing on the scientific advice available within the government. The great majority of competent academic scientists in the relevant fields are at least skeptical about the ultimate technical feasibility of the program's announced

12. See Everett Carl Ladd, Jr., and Seymour Martin Lipset, *The Divided Academy* (New York: McGraw-Hill, 1975), chap. 2 passim, esp. tab. 4, pp. 29-31.

13. Ibid., chap. 3 passim, esp. tab. 10, p. 60.

14. The first example, relating to public key cryptography, is discussed at length in Silvia Sanders, "Data Privacy," *Reason,* Jan. 1981. The second, relating to an open meeting of the Society of Photo-Optical Instrumentation Engineers, to which 2700 members from 25 countries came, was first reported in *Science News,* 4 Sept. 1982, pp. 148-49, and later in *Science,* 24 Sept. 1982, pp. 1233-34.

goals, and they strongly doubt that the announced timetables can be met.[15]

Further, the way that the program's administrators have approached the universities smacks of a mixture of bribery and coercion. Funds for academic research under the Strategic Defense Initiative Office increased sharply from 1984 to 1987, while other DoD funding for this purpose declined. In addition, the Strategic Defense Initiative Office has been very receptive to the shifting of ongoing research from other sponsors to itself as a funding source.[16] High-ranking officials in the defense research establishment have publicly expressed anger at academic criticism of the program and have threatened to cut off other defense research funding for the critics.[17] Yet, despite these anfractuosities, the relationship goes on.

15. See, for example, *The Science and Technology of Directed Energy Weapons* (New York: American Physical Society Study Group, 1987); "Weapons in Space," *Daedalus,* Spring 1985, vol. 1; U.S. Office of Technology Assessment, *Ballistic Missile Defense,* 1985; Harold Brown, "Is SDI Technically Feasible?" *Foreign Affairs,* 64(3) (1986); Sidney D. Prell, Philip J. Farley, and David Holloway, *The Reagan Strategic Defense Initiative: A Technical, Political, and Arms Control Assessment* (Stanford, CA: Stanford University, Center for International Security and Arms Control, 1984). See also Gregg Herken, "The Earthly Origins of Star Wars," *Bulletin of the Atomic Scientists,* Oct. 1987, pp. 20-28.

16. Evidence of this attitude is reported by MIT faculty. In academic year 1985-86, I chaired an ad hoc committee of the faculty on the military presence at MIT. Among other things, the committee sent a questionnaire to all members of the faculty, asking a variety of questions about sources of faculty members' research support and their attitudes toward it. A copy of the committee report is available through the Office of the Secretary of the Faculty at MIT.

17. See the statements of Donald Hicks, under secretary of defense for research and engineering, reported in *Science,* 25 Apr. 1986, p. 44.

THE FUTURE
OF THE RELATIONSHIP

Can both sides aspire to return to the golden age of the first two postwar decades? Should they?

It appears unlikely that the past can be recovered, here as elsewhere. Changes in the environment on both the defense and the university sides that will not be reversed soon, if at all, will shape the future of their relations. First, the generation that experienced the golden age is passing. Especially in the universities, the attitude that science should gladly accept an obligation to contribute to the defense establishment is vanishing with the scientists who carried it over from their World War II experience. Second, while in the first decades after the war, confrontation with the Soviet Union seemed unavoidable to the majority in the academic world and the involvement in it of a substantial military element seemed equally unavoidable, neither part of this proposition now holds. Confrontation with the Soviets no longer seems either natural or necessary to many both inside and outside the academy. Finally, the rapid and baneful evolution of the technologies of war makes the utility of military force in U.S.-Soviet relations and, indeed, in many other political conflicts around the world much less clear, although not clearly dispensable.

On the other side, the defense establishment no longer sees university scientists as the bearers of unique and indispensable skills. The FFRDCs, the laboratories of the major defense contractors and other high-technology firms, as well as those within the defense establishment now muster a formidable force of scientists and engineers more directly attuned to that establishment's needs and purposes.

Thus the basis for a sense of mutual shared interests has shrunk. Further, the relative decline of the importance of defense funding for academic science has contributed to the effects of the change in attitudes. So have the continued polarization of American politics away from a shared centrist consensus, and the consequent disjunction in the spectrum of academic and governmental views. And so, further, has the fiscal stringency, which continues to press all agencies of government to dispense with nonessentials; to many, expensive pure science is one such luxury.

Nonetheless, there are strong reasons on both sides to maintain the relation and improve it to the extent possible. The extra-university resources of scientific and engineering talent may seem to the defense establishment both a sufficient replacement for those inside the universities and one much easier to work with and manage. The second is certainly true. With respect to a wide range of development and applied research the first also may well be. But almost all the most gifted scientists will continue to reside in the universities, and, in science, even ten of the competent and diligent do not make up for one of the best. So the defense establishment's interest in continuing access to the best remains strong. In our disorderly but effective system for funding pure science, the universities have benefited greatly from the diversity of funding sources. The loss or diminution of any substantial funding source is difficult to make up, so maintaining what diversity there is remains a strong institutional interest for the universities and also tends to be so for those of the faculty who experience its benefits.

Finally, there is a broader public interest to consider. The public's continuing interest in an effective and efficient defense establishment will be better served by one that is open to both the scientific expertise and the informed criticism of the university world. Openness to the second is hardly possible without the contribution of the first. Moreover, the rapid evolution of military technologies has led to a much broader and more rapid interplay between technology and strategy in the last half century, and the relative contribution of analytical thought and imagination as against past military experience to understanding the interplay has moved in favor of analysis and imagination. By so much greater, therefore, becomes the virtue of avoiding a purely encapsulated defense establishment, talking only to itself and those who live with it every day.

Can what is needed to maintain the relations at an effective level of civility be achieved? For the defense establishment, this achievement requires three things. First and most important is to adhere to the rules of the implicit contract. Second is to look more at what the academy does than to listen to what some of its more vocal members say. Third is to be readier to hear criticism. For academic scientists, it is necessary to remember three things: first, they share the public interest in an effective and efficient defense and should respect the variety of ways that the university world can contribute to it; second, no one has a monopoly on virtuous intent, and the effective path of virtue is never easy to discern; and third, *pecunia non olet*.

The scriptural injunction that it is more blessed to give than to receive is not usually taken as applying to unsolicited advice.

Consequential Controversies

By DAVID A. WILSON

ABSTRACT: Academic science in certain disciplines and the military have had a curious quasi partnership for more than forty years. While the operators of this partnership share elements of a scientific ethos, their institutional frameworks are radically different. A number of controversies related to the relationship between the military and the university that have affected the universities, the science programs of the military departments, and the conditions of research are explored. Still the partnership persists.

David A. Wilson is professor of political science at the University of California, Los Angeles. Recently he has published on federal controls of scientific information. He is working on a study of military support of academic science at present.

DURING World War II, the military services, with the help of academic scientists, grasped the power of science in the discovery, development, and construction of a number of weapons (atomic bomb), weapons systems (hunter-killer groups), weapons system components (radar), and weapons components (proximity fuze) that substantially contributed to the effectiveness of combat forces. In the course of the war, the military also developed ways of dealing with universities that laid some foundation stones for a structure of complicated relationships between scientists in universities and the military organizations.

Following the war many, if not most, scientists working in military labs returned to or joined academia. At the same time the Navy established the Office of Naval Research (ONR), which became the pathfinder in many ways for the modes and processes of work in universities sponsored by the military. ONR developed the research contract, proposal procedures, peer or merit review, funding patterns, valuative procedures, and other operational modes that have persisted to the present in the service and Department of Defense (DoD) research organizations.[1] Many of the procedures developed by the military have been adapted for use by other federal agencies.

These accomplishments were remarkable because they permitted the flow of resources to science while sustaining substantial freedom for the investigator. They also permitted continuation of the American combination of research and graduate education. Moreover, the talents of academic scientists and engineers remained accessible to the military, presumably in ways not otherwise possible.

On the face of it, universities and military organizations might not seem compatible. Universities intend to be communities of freedom for students, scholars, and scientists while the military exalts discipline, patriotism, and martial virtues. Predictably, over four decades the complex relationships between these two groups of institutions have been frequently strained by controversy. As Price said, "The scientists as a class always seemed to be quarreling with military officers as a class, even while they were developing one of the most effective partnerships in history."[2] This article is an extended comment on this ironic remark, largely from the point of view of universities. Following some background discussion of the ethical and organizational context of the science programs, several kinds of controversies that have had consequences for universities and the DoD science agencies will be reviewed.

BACKGROUND

Programs of fundamental science sponsored by defense agencies and performed in universities operate in their own unique institutional context. Within the government the programs proceed in an environment characterized by both weighty political forces and firm structures of military organization. In universities they are conducted in the setting of the American university, with its academic attitudes, its various disciplinary elements, and its concerned student groups. Universities are also open to many political influences.

---

1. ONR was established by act of Congress in 1946, the Army Research Office in 1951, the Air Force Office of Scientific Research in 1952, and the Defense Advanced Research Projects Agency in 1958.

2. Don K. Price, *Government and Science* (New York: New York University Press, 1954), p. 57.

Scientific ethos

The effort has been sustained and defended by men and women guided significantly by the ethos of science.[3] The principles incorporated in this ethos are shared by the technical personnel of the DoD agencies and the faculty investigators in universities. Important principles of this ethos are:

1. Research problems are best defined by the state of knowledge rather than by mission objective or social goals.
2. Research is best conducted openly with open distribution of data and results.
3. Scientific merit is best determined by competitive procedures and assessment by scientists.
4. Scientific research is politically neutral and should be insulated from partisan or policy controversy.[4]

The research agenda and programs have, as might be expected, been substantially determined by scientific considerations and criteria. They have been applied, however, within constraints introduced from the institutional environment and therefore the agenda has also been shaped

3. There is a dispute, or at least discussion, about the notions of an ethos of science, but there seems to be agreement that scientists do adhere to some general principles, one way or another. Cf. William W. Lowrance, *Modern Science and Human Values* (New York: Oxford University Press, 1985), pp. 46-47.
4. Science and scientists have played their role in the tension, even conflict, in American political history between the elitist and populist poles. See, for example, Robert Post, "Science, Public Policy, and Popular Precepts: Alexander Dallas Bache and Alfred Beach as Symbolic Adversaries," in *The Sciences in the American Context: New Perspectives,* ed. Nathan Reingold (Washington, DC: Smithsonian Institution Press, 1979). Scientists' elitist position has been interpreted as harmonious with the military search for a degree of political autonomy. See David Dickson, *The New Politics of Science* (New York: Pantheon Books, 1984), p. 115.

by both federal and academic policies as well as service department missions. Tensions have predictably arisen because of pressure for more applied work. In addition, conflict over strategic policy, over war fighting, over campus turbulence, and over handling of sensitive information has influenced the conduct of these programs.

Institutional framework:
Universities

Like so many other major social activities in the United States, the conduct of scientific research is performed in a mixed and pluralistic institutional structure. In the history of the country, a federal government capacity to deal with that seeming chaos was slow in coming.[5] In the past four decades, scientists and administrators in universities and government have developed a variety of forms to permit cooperation in science.

Universities have developed flexible organizations that are able to adapt to requirements of sponsored research. Scientists—and other scholars—as faculty members are organized in departments that are defined by the disciplines or sciences—for example, physics, mathematics, chemical engineering—and are responsible for the conduct of instructional programs. Much science is done within the departmental framework. Departmental research embeds scientists and projects in a teaching organization that combines a variety of needs and talents with reasonable effectiveness.

Teaching provides a kind of subsidy or at least financial stability for faculty scientists because regular faculty compen-

5. See A. Hunter Dupree, *Science in the Federal Government* (Baltimore, MD: Johns Hopkins University Press, 1986); Daniel J. Kevles, *The Physicists* (New York: Knopf, 1977, 1978).

sation is funded from revenues derived mainly from undergraduate teaching. Moreover, the combination of graduate education with research involves students in the work and helps to keep costs down. The teaching program also serves as a moral anchor that strengthens the academic side of the relationship because arguments founded in the educational needs of students carry a special authority in the world of science.

In addition, universities have developed a variety of organized research units that are focused on more specialized areas of research interest, particularly those that require the participation of several disciplines—for example, oceanography, agriculture, medical engineering—or that require special facilities, such as astronomical observatories, accelerators, and shaking tables. Organized research was extensively developed after World War II largely in response to requirements of substantial external support, particularly federal. These institutes and centers permit team research including multidisciplinary work in units with management capacity to focus effort on a complex subject, field, or problem. Organized research units provide a degree of administrative autonomy, responsibility, and cost control in the management of externally sponsored research. They employ research scientists but mainly depend on the part-time participation of regular faculty in their research roles. Such units typically do not conduct instructional programs, although students participate as research assistants and by means of postdoctoral appointments.

Across this organizational grid is laid the ever changing mixture of projects, sponsored—from external sources—or not. Projects are led by principal investigators, normally regular departmental faculty members, who assume adminis-

trative responsibility for the conduct of the project's task in the name of the university with the support of staff employed by the university and in facilities owned by the university. Sponsored projects are governed by university policies, frequently developed in response to public policy, and the terms set forth in a funding instrument—contract or grant—which is accepted by the university in its corporate role.

Universities also possess or manage organizations dedicated substantially or entirely to research. These are variously categorized, but the distinctions that are important are between organizations that are part of a university performing work on contract from one or several federal agencies and those that are the property of the federal government and managed on contract by a university. There are some organizations in which more than one university is institutionally involved in the management.

Institutional framework: DoD

Within the DoD organization, there are a number of units that are the principal substantial sponsors of basic research in the scientific community, one for each service and one for the secretary of defense.[6] They are the Navy's ONR, the Air Force Office of Scientific Research, the Army Research Office, and the Defense Advanced Research Projects Agency. Each is located within the unique

6. Within the Office of the Strategic Defense Initiative (SDIO) there is a division of Innovative Science and Technology that sponsors university research. As an element of the SDIO, it is by budgetary definition engaged in development (budget category 6.3+) work, but much of the activity, at least as far as university performers are concerned, is clearly basic research. The National Security Agency sponsors basic research in mathematics.

structure of research and development (R&D) administration of its parent organization, and each operates somewhat differently. They are staffed, for the most part, by civilian scientists and engineers. It is these people who develop the external research programs and write contracts for university-based projects. In addition, the DoD and the services have a number of standing scientific advisory committees, consisting of persons appointed from industry and academia, that provide studies and consultation on the needs and prospects in science and technology. The Defense Science Board is the archetype of this pattern. The science agencies share the task of stimulating and sustaining a productive relationship between the military establishment and the scientific community.

To accomplish this task is as complex, if not as massive, as the somewhat parallel task of relating the military to industry. Fundamental science is largely based in universities and practiced by tenured professors motivated by a passion for discovery and ambition for scientific fame. Their professional standards are largely set by the community of their discipline and implemented by academic departments and learned societies. For the military to penetrate this dense thicket in a fashion that stimulates cooperation, let alone support, requires the service organizations to accept, even develop, procedures and practices that are open and collegial.

Universities and scientists

The quasi-autonomous status of faculty members within universities as they play their roles as teachers and as professional scientists is a structured element of the relationship between academic science and the military. Faculty members initiate

the projects that are proposed for military support. They also adopt stands on matters of university and national policy, based on their own professional ethic or political opinion. Faculty members may be heard on any matter, their professional associations speak from time to time on questions of ethics or science, but universities try to stick to those issues that affect conditions for teaching and research on campus. As institutions, universities, generally speaking, strive to maintain conditions that encourage productive endeavor of the best attainable quality. Universities have avoided the political arguments. They accept contracts and grants to support research proposed by members of faculties so long as the work complies with law, the conditions are acceptable, and the funding is adequate. Thus the burden of any political positioning lies with the investigators rather than the administrators.

CONSEQUENTIAL CONTROVERSIES

Controversies frequently have enlivened the environment of the continuing science programs. Such controversies have involved matters of strategic policy, university complicity in war, and procedures for the conduct of science. In their course, they have affected the extent of the science programs and the manner of their conduct.

The arms debate

The nationwide public disputes about strategic nuclear weapons, peripheral wars, and central front—or North Atlantic Treaty Organization—strategy have had a firm and stable connection with the academic world, although universities as institutions have not been significant parties. Together with nonprofit research corporations, universities provide a secure base for intellectuals of military matters—the majority of whom are physical and social scientists—from which they participate in the great debate over national policy regarding nuclear weapons, arms control, and strategy toward the USSR in Europe and the Third World. In one form or another, this debate has been joined since World War II.[7] For the most part, the debate has focused on strategic issues including arms control rather than involving pacifist opposition to military preparation. The controversies were most important when they publicly pitted insiders against each other on fundamental issues.

The recurrent debate over strategic defense exemplifies in the clearest way the character of these disputes. Within it are elements of strategy, technology, organizational interests, and the politics of Congress and the presidency in all its partisan richness. On this issue academic scientists have acted in important roles.

Antiballistic missiles

In 1967, the development and deployment of a system of antiballistic missiles (ABMs) for the defense of American cities was proposed by the Johnson administration.[8] Public opposition to the

7. See Fred Kaplan, *The Wizards of Armageddon* (New York: Simon & Schuster, 1983); R. A. Levine, *The Arms Debate* (Cambridge, MA: Harvard University Press, 1963). Levine is writing a review of the same issues, 25 years later, with the title "Still the Arms Debate."

8. Kaplan, *Wizards of Armageddon*, pp. 343-55. In fact, the debate started in the late 1950s but failed to attract much public attention even though the scientific community was doubtful about and, for the most part, opposed to the deployment of antiballistic missiles from the start. Secretary of Defense McNamara was skeptical. His proposal of the system was very much the result of political pressure from the military and congressional friends.

proposal from the most distinguished scientific sources quickly appeared. Hans Bethe, a theoretical physicist at Cornell University who had headed the theoretical division of the Manhattan Project, and Richard Garwin, an IBM scientist and member of the President's Science Advisory Committee and the Defense Science Board, published an attack on the idea based on both technical and strategic considerations.[9] They argued that such a system would be unreliable, relatively easy and cheap to overcome, and, should it appear to be effective in defense of American cities, very threatening and strategically destabilizing. These arguments were then and have continued to be the standard fare of the debate. The remarkable thing was publication in a popular magazine of the opposition of establishment scientists par excellence.

The controversy was emphatically reinforced a year later, in the early months of the Nixon administration, by an elaborate public hearing of the Senate Foreign Relations Committee at which Bethe, Garwin, three former presidential science advisers, and a number of other distinguished scientists from the Massachusetts Institute of Technology (MIT), Harvard, and Stanford testified to the weaknesses, uselessness, and risks of an ABM system.[10] A rather bitter struggle ensued among scientists and engineers in which the opposition to ABM was publicly attacked on technical grounds.[11] The first

9. Hans A. Bethe and Richard Garwin, "Anti Ballistic Missile System," *Scientific American,* Mar. 1968.

10. "ABM: Scientists' Loyal Opposition Finds a Forum," *Science,* 21 Mar. 1969, pp. 1309-11. The Nixon administration changed the proposed system to the defense of intercontinental ballistic missiles instead of cities, which removed the argument that the system was strategically destabilizing.

11. Kaplan, *Wizards of Armageddon,* pp. 351-55.

phase of this debate of the scientific establishment with the military was ended in 1972 with the ABM Treaty, which effectively foreclosed the idea of defense against strategic missile attack.[12]

Strategic Defense Initiative

The second major phase of the debate was initiated by President Reagan's announcement of his Strategic Defense Initiative (SDI) in 1983.[13] While the strategic and political issue, complicated by the ABM Treaty, remained essentially the same, the technological questions were different. In an important way the rehabilitation plan of a defensive strategy depended on the credibility of new technology, so the role of the scientists was truly crucial. Bethe, with Garwin and others, again took a leading role by publishing "Space Based Ballistic Missile Defense,"[14] in which they argued that the envisioned systems are perhaps impossible, certainly impractical, and probably provocative of "a large increase in Russian strategic offensive forces." Others, notably Edward Teller of the Hoover Institute, argued that feasibility could not be ruled out while the desirability of a workable defensive strategy justified the effort.[15]

The SDI had a substantially different relationship to the scientific and engineering community—academic as well as industrial—from that of the ABM effort of the 1960s. It was dependent upon systems that were little more than concepts and involved many technologies that were still very much confined to labs, if not the backs of envelopes. Thus the program as it was established in the Strategic Defense Initiative Office of the DoD was research and development, largely research. A directorate—Innovative Science and Technology—was set up with a particular mission to reach out to universities for research. Thus a substantial amount of money was available for university scientists and engineers. Their interest was solicited and there was positive response.[16] The controversy was to be played out partly in the workplace.

Scientists concerned about SDI took up the debate by organizing a boycott of SDI-sponsored research.[17] Their arguments included both the technical infeasibility and the strategic effects of a massive, space-based defensive system. Such an effort, at least on the scale of this one, was unique in the history of the arms debate. A significant number of academic scientists and engineers were personally and collectively putting their participation in research on the table. An intention of political intervention in the arms debate was clear in the exploitation of the public relations potential and the solicitation of congressional endorsements by the boycott organizers. Proponents of SDI in the DoD made substantial efforts to belittle the boycott, but it would be difficult to deny the contribution of this activity to the character of the national policy discussion. A technique was exhibited here that related the performance of academic research to the issue of end use through the voluntary choice of scientists without

12. See Joel Primack and Frank von Hippel, *Advice and Dissent: Scientists in the Political Arena* (New York: Basic Books, 1974), pp. 59-73, 178-95.

13. "Address to the Nation on the Strategic Defense Initiative," 23 Mar. 1983.

14. *Scientific American*, Oct. 1984, pp. 39-49.

15. "SDI: The Last, Best Hope," reprinted from *Insight* in *Strategic Defense Initiative*, ed. P. Edward Haley and Jack Merritt (Boulder, CO: Westview Press, 1986).

16. "Star Wars Grants Attract Universities," *Science*, 19 Apr. 1985, p. 304.

17. John Kogut and Michael Weissman, "Taking the Pledge against Star Wars," *Bulletin of the Atomic Scientists*, Jan. 1986, pp. 27-30.

enlisting universities as institutions. Universities were not constrained to a position on national policy or required to substitute a corporate judgment for the judgment of professional scientists. By this means, a persistent dilemma for universities was avoided by the opponents of SDI.[18]

University "complicity" in war

The campus turbulence during the period from 1964 to 1972 was focused in large measure on issues arising from the Vietnam war. In these disputes the role of universities as institutions in the draft, the Reserve Officers' Training Corps (ROTC), and war research was salient.

Since World War II, many universities had accumulated an array of research and public service programs sponsored by federal agencies, notably but by no means exclusively the DoD.[19] In the atmosphere of excitement and urgency there was little consideration of any effect on the character of the institutions, nor were policies to govern these activities very much in evidence.[20] In the late 1960s a number of institutions, under various

degrees of pressure, undertook reviews of their situation as related to sponsored "war-related" research and other service programs. Learned societies, mostly in the social sciences, were also exercised by questions of scholarly participation in activities related to the war. These questions were characterized as a matter of "professional ethics."[21]

The significance of debates about "war research" was not only the passion, fear, and anger that was demonstrated in the late 1960s, all of which seized the attention of most people on campuses and otherwise interested in universities, but also the revelation of a broad range of opinions about externally sponsored activity. Importance was to be found in the lasting consequences. For the question of the relationship between the university community and the military, particularly in science, the most significant consequence followed from a chaotic but genuine review and rethinking of the terms of the deal.[22] Both universities and the agencies of government found an occasion for this sort of broken-backed review of what kind of work under what conditions was appropriate.

The controversy over "war research" at universities was focused on a variety of observed or imputed consequences of

18. The proponents initially sought to enlist the names of institutions by the creation of putative consortia of universities when faculty investigators were working on SDI contracts. After complaints from university presidents, this practice was abandoned.

19. Clark Kerr, then president of the University of California, published his 1963 Godkin lectures as *The Uses of the University* (Cambridge, MA: Harvard University Press, 1963), in which he discussed in one lecture "the realities of the federal grant university." In it he pointed out that 75 percent of all university expenditure on research was federally funded, of which 32 percent was from the DoD. He made a number of observations about the consequences for internal cohesion and control of this situation.

20. Agency for International Development programs in South Vietnam and Thailand, the interest

of Project Jason in counterinsurgency in Southeast Asia, and the Advanced Research Projects Agency's extension of its programs into the social sciences were particularly notable instances of programs involving universities and faculty that precipitated impassioned protests during the latter part of the 1960s. See, for example, Seymour J. Deitchman, *The Best-Laid Schemes: A Tale of Social Research and Bureaucracy* (Cambridge: MIT Press, 1976).

21. Harold Orlans, *Contracting for Knowledge* (San Francisco: Jossey-Bass, 1973), pp. 15-80.

22. See Carl Kaysen, "Can Universities Cooperate with the Defense Establishment?" this issue of *The Annals* of the American Academy of Political and Social Science.

scientific and engineering effort somehow related to the military. Such undesirable consequences may be separated into two categories: (1) corruption of science and the university in some way, and (2) deleterious effects on society. "Corruption of science" referred to such matters as secrecy, effort devoted to applied research and product development rather than to the pursuit of new knowledge, and the subordination of academic work to military need rather than scientific purpose. Deleterious social effects were to be found in perceived contributions of research to oppression and destruction rather than to freedom and equity.

Debate and discussion of these issues, which often took place in an environment of forceful if not violent student demonstrations, were heated. Some resolution was possible in those years, but this same argument persists to the present. It centers on the questions of the academic and scientific effect and the environmental and social impact of research and service projects undertaken by university faculty. Answers to these questions, it is argued, can provide standards of appropriateness by which choices between projects can be made. Depending on how such standards are defined, there may be within them the potentiality of affecting the programs of both the universities and the military research agencies as well as the relationship between them. Universities are wrestling with the matter of appropriateness in a variety of contexts.

The difficulties are these. Within the very broad and flexible range adumbrated by the conventional academic missions of teaching, research, and public service, the activities of universities are largely driven by faculty initiative and effort. The corporate entity of the university has an important but still peripheral role in providing guidance and constraints for faculty activities. A definition of what is appropriate to a university could provide a critical constraint on the substance of research and service programs.

It has proven to be possible to develop consensus on standards related to academic and scientific integrity since such integrity seems to be an elemental part of the moral core of the institution. The logic of the consensus springs from a commitment to teach the truth unbiased by sectarian or political considerations, to seek in research the truth and, particularly in science, new knowledge about the phenomenal world, and to conduct research in a manner compatible with the obligations of teaching, thus openly and also free from sectarian or political bias. This logic has helped to define integrity and to assess academic and scientific impact.

At the same time, however, it has rendered standards derived from questions of social impact ineffectual and unacceptable. Such considerations very quickly carry the issue into areas that are clearly matters of public policy about which substantial and usually lively differences of opinion prevail, on campus and off, now as well as two decades ago. The resolution of the question of appropriateness by this route clearly would be a matter of politics, itself ruled out of order. For this reason, universities have adopted stands of institutional neutrality explicitly eschewing public policy positions. These stands seem somewhat ambivalent, particularly because deeply embedded in the ethos of universities there are unresolved conflicts between teaching and research, pure and practical science, basic and applied research, research and public service. In principle—that is, in the abstract—this conflict is always resolved in favor of the educational and the pure even though the immediate and practical persist. Just so in the rejection of arguments about standards of social and policy position even while effort with clear practical consequences goes on.

The draft

While selective service was not administered by the military departments, in the latter part of the 1960s in the context of the Vietnam war, it was depicted as a coercive process that used the threat of military duty to constrain the behavior of young men. Undergraduate students and graduate students in certain fields who were in good standing had their service deferred. This situation necessarily involved universities as institutions and their faculty in a set of impassioned relationships that were judged by some to be matters of life and death. Grades, attendance, admission, and enrollment matters affected the fates of draft-eligible students. Faculty and administrators were embroiled in tense personal confrontations with students in which their life expectancy, not merely their career choices, was presumed to be at issue. Particular matters that compounded the problem were the requirement of selective service that universities provide reports of student performance to draft boards and also the continuation of ROTC programs. As might be expected, such tension generated a great deal of anxiety and resistance to the presumptive "complicity" of universities in the war effort.

Responding to the pressure of campus turbulence, some universities discontinued the practice of reports to draft boards except at the request of students.[23] With the end of the war and conscription, the issue went away.

ROTC

As for ROTC, the matter followed a more academic course. The program was treated by universities as a department, offering courses in military science and staffed by military officers assimilated to appropriate academic titles. The program is mandated to higher-educational institutions by statutes whose lineage is in the first Morrill Act establishing land-grant colleges. Universities in this category felt themselves to be not free to abolish ROTC altogether as was demanded by the protest movements.

As was the case generally with student demands, the issue of ROTC took on effective force within universities when a significant part of the faculty took up the cause. The program came under scrutiny from faculty senates on the question of its academic compatibility. The faculty status of officers assigned to campuses as well as the granting of credit for military science courses became the focus of dispute at many institutions. In more than one institution, faculty decided in heated, emotional, and often very pressured debate to discontinue academic credit for military science courses on the grounds that they were "purely professional" and to discontinue professorial appointments for officers, largely on the grounds that these appointments did not meet the normal standards for the professorship.[24] Lengthy negotiations with the services followed that, at least in the case of some private institutions, resulted in discontinuing the programs. In others, agreements were reached that brought the programs under greater faculty supervision.[25]

23. Immanuel Wallerstein and Paul Starr, *The University Crisis Reader* (New York: Random House, 1971), 1:208-9.

24. Ibid., pp. 262-91.
25. See Richard M. Abrams, "The U.S. Military and Higher Education: A Brief History," this issue of *The Annals* of the American Academy of

War research

Much of the energy of protest was directed to attacks on "war research." The events of the late 1960s exposed the latent conflicts and internal tensions arising from the characteristics of federal, particularly defense, programs, related to balance between disciplinary areas, the management of major laboratories and facilities, and the openness and ethics of research activities.

Concerns about war research were manifested in various events in these years, but, perhaps, the convocations of 4 March 1969 and what followed were the most revealing of the structure of the situation.[26] Apparently at the initiative of graduate students at MIT, a number of convocations were held on various campuses around the country to focus attention on the performance of research related, in some way, to the military and the war in Vietnam.[27] These convocations revealed a spectrum of views ranging from the radicals who sought cessation or "reconversion," the liberals who sought greater openness and consideration of appropriateness, and the conservatives who, although not vocal, presumably sought to maintain the status quo ante.[28] The administrations of the universities involved were principally concerned about the effect of these activities on the institution. For the most part, administration spokesmen eschewed comment on national policy questions, sticking with issues that were largely matters of "time, place, and manner" such as the use of force, the sanctity of contracts, the need for openness, and the centrality of free speech and academic freedom.

The pressured review of research had four kinds of consequences, three for university administration and one for the DoD. A number of universities separated themselves from organizations that conducted substantial defense-related research. The most conspicuous of these actions was Stanford's detachment of the Stanford Research Institute, Cornell's spin-off of the Cornell Aeronautical Laboratory, and MIT's separation of the Instrumentation (Draper) Laboratory. These actions were not without pain and cost. To varying degrees these enterprises involved the academic programs of the university, which had to be disentangled, and generated revenues that were lost and were associated with notable scientists who were offended. The effect, however, was not catastrophic. Other universities retained their management of government-owned laboratories in spite of pressure from the campus communities to sever ties or to reconvert to non-weapons-related research.[29]

Second, the review of war research on campus came to focus on the question of secrecy. Many universities were hosts to federally sponsored projects that involved some degree of official classification. The extent of these projects throughout the academic community is obscure because data are elusive or nonexistent. Secrecy was an issue that could generate broad agreement in the conflict-ridden environment of the time. Freedom and openness

Political and Social Science; on analogous issues in the USSR, see Julian Cooper, "The Military and Higher Education in the USSR," ibid.

26. *Science,* 14 Mar. 1969, pp. 1175-78.

27. While work sponsored by military agencies was under scrutiny and attack, research and service contracts from the Central Intelligence Agency, the Agency for International Development, and other agencies were also significant.

28. At MIT the radicals organized the Science Action Coordinating Committee while the liberals established the Union of Concerned Scientists;

29. California, Johns Hopkins, and Rochester all kept relationships with military laboratories.

have been greatly esteemed characteristics of the academic community since the earliest years many centuries ago. Struggles over these matters are part of the lore. Classification by the government of information and of results of research was antithetical to these deeply and broadly held values even though values of service and patriotism were countervailing. The matter was particularly unseemly in the case of graduate students who, as a result of their participation in classified research projects, ended up with secret dissertations. The existence of such things was surprising and shocking to students and, more significant, to rather broad parts of faculties.

At university after university, committees were formed to review the matter, and policies were formulated and adopted to forbid or to closely regulate classified research on campus. Counterarguments that such regulatory policies constituted a violation of academic freedom were not effective.[30] Such policies were carefully drawn to avoid absolute barriers by such distinctions as those between on- and off-campus locations or between receiving classified information as opposed to generating it in the course of research. But, for the most part, disapproval of graduate-student dissertations that were classified and withdrawal from classified work was the rule.[31]

Finally, universities began to take a more serious and focused view of the appropriateness of extramurally supported projects. Such matters are fundamentally judgmental, and therefore, as far as universities were concerned, they led to procedural safeguards involving peer reviews. For example, at MIT the Sheehan committee was set up to evaluate, among other things, the acceptability of specific contract proposals at the Special Laboratories.[32] At Michigan, a committee was established to evaluate proposed classified contracts. At California, the Zinner committee reviewed the place of the Atomic Energy Commission laboratories in the university. But criteria of potential end use were too contentious to be effective. For the most part, project substance failed as a standard of appropriateness.

Mission-related research

Scandals about some projects, especially the pitiful Project Camelot,[33] aroused congressional interest in the question. Senator William Fulbright took the position that universities were risking their role as "idealistic rather than pragmatic" institutions committed "to moral rather than expedient purposes" as they edged closer to becoming "a paid producer of goods and services, a hireling of the state."[34] Fulbright offered an amendment to the military procurement authorization for 1970 to restrict research support to projects or studies that had "direct and apparent relationship to a specific military function or operation." It was adopted and became known as the Mans-

30. Wallerstein and Starr, *University Crisis Reader,* pp. 221-37.

31. The move to regulate secrecy on campus was an early example of a line of regulatory requirements for the conditions of research generated by both university and government concerns. Others of importance are the use of human subjects, the use of live animal subjects, biological safety, conflict of interest, and, often, standards of appropriateness for the academic community.

32. Dorothy Nelkin, *The University and Military Research* (Ithaca, NY: Cornell University Press, 1972), p. 88.

33. See Deitchman, *Best-Laid Schemes;* Irving Louis Horowitz, *The Rise and Fall of Project Camelot,* rev. ed. (Cambridge: MIT Press, 1974).

34. Speech at Denison University, 18 Apr. 1969, reprinted in part in Wallerstein and Starr, *University Crisis Reader,* pp. 237-38.

field Amendment[35] because of the support of Mike Mansfield, then majority leader of the Senate.[36] Thus the effect on DoD agencies of the pressure to review was to require that they also face the question of appropriateness.

There is an implicit and persistent tension in the relationship where the direction of research effort purports to derive from either a motive of discovery in an arena defined by the state of knowledge or a motive of technological advance in an arena defined by mission need.[37] Management of the tension is referred to as the user-supplier dialogue or coupling.[38] From the perspective of this article, the key coupling is between the academic community and the science agencies of the military. Couplings involving military labs and industrial performers are significant in the R&D process but of limited effect on the course of academic science.

From within the services and the administrative processes of the DoD there have been expectations and pressures to demonstrate the contribution of research activities to the cause. Since the passage of the Mansfield Amendment, research agencies have been sensitive to congressional scrutiny as well. Considerable effort is expended to justify, rationalize, and explicate the basic science programs in relation to the military mission of the services and the DoD. This has been accomplished by establishing technology-related elements of the programs, such as aerospace sciences; by designation of objectives that link scientific issues with service needs, such as "synthesis of novel inorganic, organic and organometallic compounds of special naval interest"; and by focused programs, such as programs concerning energy-absorption phenomena "targeted at making possible development of low-observable structures."[39]

35. Pub. L. 91-121, 203.

36. Rodney W. Nichols, "Mission-Oriented R&D," *Science,* 2 Apr. 1971, pp. 29-37. In the authorization for 1971, the language was changed to the much less restrictive "in the opinion of the Secretary of Defense, a potential relationship [to a military function or operation]." The congressional motive for this provision was surely mixed, but the effect was to require the department to review its research program and reflect upon its relationship to the agency's mission, which is itself rather complex, to say the least.

37. The defense establishment conceptualizes the process of research and development as a continuous movement from fundamental knowledge to a working prototype ready for production. The subcategories of the DoD R&D budget reify that conceptual scheme as follows: 6.1, research; 6.2, exploratory development; 6.3, advanced development; 6.4, engineering development; 6.5, management and support; and 6.6, operational systems development. About 10 percent of the R&D budget, which includes research (6.1), exploratory development (6.2), and concept feasibility (6.3a), comprises the so-called technology base. University-based programs are overwhelmingly within the technology-base area.

38. U.S. Department of Defense, Department of the Navy, *RDT&E: Acquisition Management Guide,* 10th ed., Jan. 1987, pp. 2.7-2.8. Some hint of the problem of coupling is to be found in the following from a news story by Grace Shinamoto of the Ottaway News Service, published in the *Sentinel* (Santa Cruz, CA), 20 Mar. 1988:

"The Office of Naval Research, founded in 1946 to carry on the scientific and technological progress fueled by defense needs during World War II, is spending nearly $1.17 million in Santa Cruz this year.

"'They sponsor research without knowing what will come out of it because the better they know their environment, the better they'll know how to operate in it,' said Theodore Foster, a UCSC marine sciences professor who is using a $66,498 ONR contract to study circulation of ocean waters.

"But the ONR spokesmen say the office funds projects it anticipates will yield technology useful to the fleet, usually within five to 10 years.

"'We are the Navy after all. We're part of the Defense Department,' said ONR spokesman Marc Whetstone." Who is right?

39. U.S. Department of Defense, Department of the Navy, Office of Naval Research, *Guide to Programs,* July 1987, pp. 10, 28.

At the same time, announcements to the scientific community call for "unsolicited" proposals within the areas described in the announcements. Emphasis in selection is laid on scientific merit and novelty of concept. Work is conducted under the direction of the principal investigator in academic conditions that include the participation of graduate students. The projects are as pure, basic, or applied as the investigator feels appropriate.

Selection is also subject to a criterion of DoD relevance.[40] Relevance is established by a complex of determinations that produce guidance for the research offices of the DoD. Programs are developed in accordance with this guidance, and projects are ultimately linked through the programs to service requirements. Documentation is produced by the technical officer for each project proposed for funding.[41] While there is clearly interaction between technical officers and faculty investigators,[42] it is not a surprise, given their similarity of ethos, that there is no evidence of overt direction, let alone coercive behavior.

Mission relationship is thus the outcome of the definition of the elements of the program and the selection for funding of projects from among proposals submitted.[43] This process is enhanced by the use of focused programs within the larger program elements that the agencies implement. These focused programs—for example, the ONR's Accelerated Research Initiatives—provide greater funding stability and other preferences for approved projects.

Skewing of science

In addition to the moral arguments about the university's participation in war research, there has been controversy about the skewing of research by the various requirements of the DoD. The most global of these arguments, put forward in considerable detail by Holdren and Green,[44] is that the great increase in defense R&D has drawn resources away from R&D for other sectors of the economy without producing a sufficient compensating positive output. "It is certainly possible, therefore," they say,

that rearrangement of national priorities in pursuit of increased military strength—and, of particular interest here, rearrangement of R&D priorities to favor military projects—could produce the opposite of the intended effect, undermining the nonmilitary dimensions of our national security in ways that outweigh any military gains.

Their criticism applies to universities not only as losers in the competition for funds but also, because of dependence upon DoD support, as losers because of potential inhibition of independent critical analysis.

A somewhat more narrow concern is that, because of the imperatives and dominance of military funding, disciplines are skewed in significant ways. Stuart W.

40. U.S. Department of Defense, *The Department of Defense Report on the Merit Review Process for Competitive Selection of University Research Projects and an Analysis of the Potential of Expanding the Geographical Distribution of Research,* prepared for U.S. Congress, Committees on Appropriations, Apr. 1987, p. 5.

41. Stanton A. Glantz and Norm V. Albers, "Department of Defense R&D in the University," *Science,* 22 Nov. 1974, pp. 706-11.

42. Department of Defense, *Report on the Merit Review Process,* p. 5.

43. Glantz and Albers reported that in the early 1970s there were from 4 to 10 proposals submitted for each funded. Glantz and Albers, "Department of Defense R&D."

44. John Holdren and F. Bailey Green, "Military Spending, the SDI, and Government Support of Research and Development: Effects on the Economy and the Health of American Science," *FAS Public Interest Report,* Sept. 1986.

Leslie[45] concludes his history of the growth of electrical engineering after World War II with the observation:

Terman had guessed correctly that the booming postwar period would be an intensely competitive one for universities, like "football without conference rules." Stanford learned to win in that high stakes environment. But its very success raised troubling, and enduring, questions. To what extent did the military establishment set the conference rules? What difference did its intervention make in how the game was played, how the score was kept, and what winning and losing meant?[46]

Paul Forman[47] argues that military support for physics pulled the discipline into quantum electronics and solid-state physics to the detriment of other issues, pulled in the direction of applied rather than pure science, and even affected the theoretical basis of the discipline. Others share the concern without necessarily going as far for their conclusions.[48]

The question of skewing is a difficult one to resolve conclusively because it is a matter of alternative histories. Clearly, the world of American science and engineering would be different had there been no programs funded by the military services, but we cannot know what that alternative world might have been. Whether it would have been better in any way is speculative. A shortage of talent

and capital for alternative areas of research suggests some misdirection of policy. As a policy question it can be addressed, but for the most part the matter has not exercised the scientific community.[49]

*Equity and merit
in research*

Another concern of a rather different nature came from the university community itself, although it found its most effective articulation in Congress. This directed attention to the distribution of resources for university research throughout the community and the country. Both the communities of science and of universities are meritocratic. Status in science is largely a matter of achievement most clearly measured by first-published research and continued citation. Universities gain their status in the academic world largely by ratings of graduate programs, which are substantially affected by the numbers and status of scientists and scholars who are members of faculties. Politicians are, on the other hand, egalitarian. Their fame and fortune depend on the effectiveness of their representation of the broadest range of groups and interests. The two worlds have a natural dissonance, even antagonism.

One of the miracles of modern science is the way in which the community has successfully sustained a pattern for distribution of public funds for the support of research that is based on merit substantially implemented by peer review. This system was largely developed by the military agencies, particularly ONR, in the years soon after World War II and has been carried over to the National

45. "Playing the Education Game to Win," *Historical Studies in the Physical and Biological Sciences,* 1987, vol. 18, pt. 1, pp. 55-88.

46. Ibid., pp. 87-88.

47. "Behind Quantum Electronics: National Security as a Basis for Physical Research in the United States, 1940-1960," *Historical Studies in the Physical and Biological Sciences,* 1987, vol. 18, pt. 1, pp. 149-229, particularly p. 200 ff.

48. See Vera Kistiakowsky, "Military Funding of University Research," this issue of *The Annals* of the American Academy of Political and Social Science; Edward Gerjuoy and Elizabeth Urey Baranger, "The Physical Sciences and Mathematics," ibid.

49. Within advisory committees to the DoD, the issue of discipline support is occasionally considered.

Science Foundation, the National Institutes of Health, and other civilian organizations. Peer review has been significantly modified in DoD by requirements for mission relevance and also by special programs aimed toward the development of more and more widely distributed institutions. Until very recently, however, the principle of merit has retained its power against an alternative principle of equity, or, as some might phrase it, the distribution principle of the pork barrel.

The power of the merit principle rests for the most part on the pretty much universal consent of the scientific and university worlds. But in recent years, as research appropriations increased and as the financial strains of years of deferred maintenance took their toll on academic facilities, temptation became too great. Peer review was reinterpreted by some universities as the old-boy network, and merit became the Matthew effect.[50] Consensus was fractured and bits of DoD appropriations were earmarked for specific projects over the strenuous objections of both the association of leading research universities and the DoD.[51]

Control of information

Although by the 1970s universities broadly had developed policies to exclude classified research projects from campuses and thus from programs involving graduate students, the issue of restrictions on access to information and the dissemi-nation of research results arose again in the form of controls on the export of technology. The strategy of American defense depends on the technological superiority of U.S. forces. Protection of such superiority includes control of the export of weapons and also control of the export of technology and technical data with potential or actual military applications. Beginning in the late 1970s, concern about the loss of technology to the Soviet Union led to an approach that sought to control key technologies as early in their development as possible. By 1980, this effort had begun to manifest itself in the form of provisions in DoD research contracts that restricted access to research information by foreigners and restricted dissemination of data and results in forms that would be accessible to foreigners.[52]

Restrictions of this kind raised issues at the very core of science. Free and open dissemination of research results is a key tenet, not only as a matter of privilege but as a matter of obligation. Because science is a collective enterprise, advance of knowledge is cumulative and cooperative. Advance depends upon the sharing of data and findings among scientists in the several disciplines and subdisciplines by both informal and formal communication. In addition, the quality control of scientific work is based on open publi-

50. Matt. 13:12. The principle, first identified by Robert K. Merton (*The Sociology of Science*, ed. Norman Storer, [Chicago: University of Chicago Press, 1973], pp. 445-47), can be roughly framed as, "Them that has, gets."

51. Frank Clifford, "Research Funds: Not So Scientific," *Los Angeles Times,* 27 Nov. 1987; Daniel S. Greenberg, "It Was a Great Year for Pork Barrel R&D Funding," *Science and Government Report,* 1 Feb. 1988, p. 3.

52. For the most part, exports are controlled by two statutes: the Export Administration Act and the Arms Export Control Act. The latter is specific to "military articles," that is, weapons, while the former controls all other commodities, including "dual-use" technologies, that is, technology that has both civilian and military applications. The administrative processes of the two acts are different and are administered by different agencies, but they both require licensing of exports. While the law gives the DoD a strong role in this activity, the DoD administers neither act. It includes itself by providing advice to the administering agencies and adding requirements to contracts with suppliers of "sensitive" technologies and materials.

cation in refereed journals of articles subject to review and criticism by the community. Restrictions are, therefore, not merely irritating but, in fact, are judged to increase the risk of slowing advances and degrading the reliability of research.

Universities, which see their responsibilities to include the protection of the environment of science, responded to this development with vigorous opposition. These institutions also had two additional reasons to resist the imposition of such restrictions. First, many graduate students in science and engineering are foreign. Restriction of access to information would have resulted in unwieldy and, on many campuses, unacceptable discrimination among students in the pursuit of research. Second, restriction on dissemination would have involved prepublication review by DoD monitors as well as substantial inhibition of the sharing of current research data among scientists not only abroad but also throughout the country. Such restrictions are anathema to most universities, which refuse to yield control over publication to any outside sponsor.

An elaborate national discussion ensued involving the National Academy of Science, various scientific societies, the university community, and various federal agencies, notably the DoD.[53] There was considerable dispute within the government and within the DoD about the advisability of applying this sort of restrictive regulation to science. After several years, national policy emerged that in effect exempted fundamental research undertaken on university campuses, including that sponsored by the DoD, from restrictions of this kind.[54]

CONCLUSION

Universities, at least the major research universities of this country, and the DoD have a common interest. The persistence of the science programs over a period of forty years in spite of controversy on campuses, in Congress, and probably in the DoD is strong evidence that the interest and its commonality are real. Some universities, some disciplines, and some individual scientists are dependent upon support from military agencies for their scientific work. Because the services are so firmly oriented toward technical superiority and thus toward research and development as a major commitment, the reliance upon the academic community for fundamental research, for the training of scientists and engineers, and for scientific consultants is as permanent an element of their effort as anything is.

Embodying this common interest are the scientists in academic departments and in the science agencies who share an outlook—an ethos—that permits the continuation of the research programs even though they may be skeptically scrutinized, or even attacked, by academic or military colleagues. The paradox of the "effective partnership"[55] between the

53. David A. Wilson, "National Security Control of Technological Information," *Jurimetrics,* 25: 109-29 (Winter 1985); Ruth Greenstein, "National Security Controls on Scientific Information," *Jurimetrics,* 23:89 (Fall 1982); Edward Gerjuoy, "Controls on Scientific Information," *Yale Law and Policy Review,* 3:447-78 (1985); David A. Wilson, "Federal Control of Information in Academic Science," *Jurimetrics,* 27:283-96 (Spring 1987).

54. Association of American Universities, *National Security Controls and University Research: Information for Investigators and Administrators* (Washington, DC: Association of American Universities, 1987). For a somewhat less sanguine view of this matter and related issues, see John Shattuck and Muriel Morisey Spence, "When Government Controls Information," *Technology Review,* Apr. 1988, pp. 62-73.

55. See the quotation from Price at the beginning of this article.

academy and the military is more apparent than real because the relationship is conducted by those men and women who successfully survive and prosper in the large and complicated organizations of which they are a part.

The programs are not, of course, completely insulated from broader influences. The academic requirements for open research and freedom to publish and for investigator-initiated projects are accepted by the military even though vociferously security-minded and directorially inclined people may complain and sometimes momentarily intervene. At the same time, the academic scientists accept the need to define the program's scope in large part by long-run mission requirements even while complaining about directed research programs and facing down accusations of complicity in war making. But these tensions require attention. Whatever equilibrium there may be is not automatic.

University administrations, military commands, and political officials support the university-military undertaking for their various reasons. Such support combined with the skill of the operators provides what is needed to give considerable stability and productivity to this curious "partnership," unique to the American scene.

The Physical Sciences
and Mathematics

By EDWARD GERJUOY and ELIZABETH UREY BARANGER

ABSTRACT: University faculty in the physical sciences and mathematics—the quantitative sciences—must generate their own research funds, often including funds to support their graduate students. The major funding source is the federal government, through its various funding agencies; the competition for research funds is intense. As a result, an agency's funding criteria, which depend on the agency's mission, can greatly influence research directions in a broad field of quantitative science. In particular, such influences—amounting to a significant skewing of research directions and subfield growth in university quantitative science, with possible effects on university recruiting and administrative policies as well—have resulted from the mission orientation of the Department of Defense (DoD), coupled with recent increases in DoD funding. Concluding observations discuss some possible means of alleviating the problems posed by DoD funding of university quantitative science.

Edward Gerjuoy has a physics Ph.D. and a J.D. He has published many papers on various theoretical physics topics, as well as on public policy issues involving law and science. Currently he is professor of physics emeritus at the University of Pittsburgh and is of counsel to a Pittsburgh law firm.

Elizabeth Urey Baranger is currently professor of physics and dean of graduate studies of arts and sciences at the University of Pittsburgh. She has served on numerous national committees including the Panel on Public Affairs and the Council of the American Physical Society, the Executive Committee of the Association of Graduate Schools, and the Graduate Record Examination Board.

THIS article is concerned primarily with the effects on universities of recent trends in the funding of research in the physical sciences and mathematics—the quantitative sciences[1]—especially with the effects of such funding by the Department of Defense (DoD). By the "physical sciences" we mean physics, astronomy, and chemistry, consistent with National Science Foundation (NSF) usage; in "mathematics" we do not include computer science, discussed elsewhere in this issue of *The Annals*.

Also discussed elsewhere in this issue is the history of federal funding of scientific research in this nation before and during World War II. The profound military impact of U.S. quantitative science research during World War II, research initiated largely by university professors, led the Navy to set up the Office of Naval Research (ONR) on a temporary basis in May 1945, even before Japan surrendered. By the time the bill permanently establishing ONR was signed into law, in August 1946, the agency already had in force 177 contracts totaling $24 million with 81 laboratories—including private or industrial as well as university labs, however;[2] by 1949, ONR was supporting 1200 projects in 200 institutions, involving nearly 3000 quantitative scientists and 2500 graduate students.[3] In 1940, the total expenditures by colleges and universities for scientific research in all fields

had been only $31.45 million.[4] Forman estimates that in the late 1930s the total annual expenditure on research by all U.S. university physics departments, exclusive of faculty salaries, was no more than $750,000.[5]

These early ONR grants provided a vastly greater source of research funding for university physical scientists than ever before had been available, especially for competent scientists at those less prestigious institutions that before World War II had rarely attracted foundation support. ONR support also was available for university mathematics research, but in precomputer days the augmentation of salary budgets—for example, to pay the principal investigator's summer salary or to enable the retention of graduate students who were not needed as teaching assistants—was the only real function of ONR grants to university mathematicians. In addition, academic scientists, especially physicists, were able to secure grants from the Atomic Energy Commission, whose basic research-sponsorship role largely was transferred to the Department of Energy (DoE) in 1977. Furthermore, federal agency grant support was quite generously—in fact, exceptionally so compared to the years before World War II—available for scientists' travel within and outside the United States as well as for defraying the expenses of arranging scientific conferences.

Especially noteworthy is the fact that, at least initially, ONR funds for academic principal investigators were awarded only

---

1. Henceforth in this article, the phrase "the quantitative sciences" will denote "the physical sciences and mathematics," to avoid repetitive inclusion of the awkward appendage "and mathematics"; the phrase "quantitative scientists" is to be similarly interpreted.

2. Daniel J. Kevles, *The Physicists* (New York: Knopf, 1979), p. 355.

3. U.S. National Academy of Sciences, *Federal Support of Basic Research in Institutions of Higher Learning*, 1964, p. 38.

4. Vannevar Bush, *Science, the Endless Frontier* (Washington, DC: Government Printing Office, 1945), p. 80.

5. Paul Forman, "Behind Quantum Electronics: National Security as Basis for Physical Research in the United States 1940-1960," *Historical Studies in the Physical and Biological Sciences*, 1987, vol. 18, pt. 1, p. 188.

after proposal review by civilian scientific adviser peers, without restrictions on publication and without reference to military utility;[6] although the awards usually were made to the investigator's university to ease the investigator's accounting and administrative responsibilities, the funds could be spent only by the investigator at his or her sole discretion, assuming, of course, that there was adherence to the original purposes of the grant. The Army Research Office and the Air Force Office of Scientific Research, established in 1951 and 1952, respectively, operated similarly at first. Nevertheless, many university faculty members and other Americans expressed great concern about the possible domination of U.S. university science by the military.[7] These concerns ultimately led to the creation, in 1950, of the NSF, whose university grants policies— though developed wholly under civilian authority—have not been distinguishable from the corresponding ONR, Army Research Office, and Air Force Office of Scientific Research policies in those DoD agencies' early years.

Grants to individual principal investigators have not been the government's only mechanism for support of university quantitative science. In particular, the government has set up numerous Federally Funded Research and Development Centers (FFRDCs). The FFRDCs are exclusively or substantially financed by the federal government but are not part of the government; instead, each FFRDC is administered by some nongovernmental institution.[8] Many of the FFRDCs

furnish experimental facilities, such as astronomical observatories or large accelerators, to user groups based in universities, and frequently they also hire university professors as consultants or as summer employees. Relevant examples are the NSF National Radio Astronomy Observatory in West Virginia, administered by the Association of Universities for Research in Astronomy, Inc., and various so-called National Laboratories supported by the DoE.[9] A number of these DoE FFRDCs—such as the Brookhaven National Laboratory on Long Island and the Fermi National Accelerator Laboratory near Chicago, which are administered by university consortia—serve primarily as facilities where academic researchers can perform high-energy fundamental-particle physics experiments; a major function of other DoE FFRDCs, however—notably the Los Alamos and Lawrence Livermore National Laboratories, which are administered by the University of California under a fee arrangement—is nuclear weapons development, a highly secret activity.

Several FFRDCs have been or are located on or very near the campuses of the universities administering them. Such FFRDCs, which typically are predominantly staffed by untenured researchers who must find their own salary and research support but who do not wish to work for government or industry, enable

---

6. National Academy of Sciences, *Federal Support of Basic Research*, pp. 36-39; U.S. National Science Foundation, National Science Board, *Basic Research in the Mission Agencies*, 1978, p. 286.

7. Kevles, *Physicists*, p. 355.

8. The criteria and rationale for setting up FFRDCs are given in U.S. General Accounting

Office, *Issues on Establishing and Using Federally Funded Research and Development Centers*, GAO/NSIAD-88-22, 1988, pp. 26-29.

9. The university-managed FFRDCs and their sponsoring agencies are listed in *Federal Support to Universities, Colleges, and Selected Nonprofit Institutions, Fiscal Year: 1986*, NSF Surveys of Science Resources Series, report NSF-318, Jan. 1988, p. 6. The 10 DoD-funded FFRDCs, only 2 of which are university managed, are listed in General Accounting Office, *Research and Development Centers*, p. 14.

a university to assemble at its campus a large pool of frequently very qualified researchers to whom the university has no permanent commitment should research funds dry up; the facilities and talents at these FFRDCs often are used for graduate-student training as well. Examples here are the Princeton Plasma Physics Laboratory, which is a DoE FFRDC; the Jet Propulsion Laboratory located at Cal Tech and funded by the National Aeronautics and Space Administration (NASA); and the DoD-supported University of Wisconsin Army Mathematics Research Center, which was, however, phased out in 1970.

The NSF budget has grown from $3.47 million in 1952 to $1.7 billion in 1988. In constant dollars, however, rapid growth of the NSF budget occurred only in the years preceding 1965. NSF research support for universities grew considerably more slowly after 1965, but it did double in constant dollars in the two decades between 1965 and 1986.[10] During these same two to three decades, DoD funding of university research fluctuated considerably. In particular, DoD support of research and development (R&D) in academic institutions almost doubled in constant dollars between 1960 and 1964, then decreased by a factor of 2.5 during the years from 1964 to 1975, after which there was another increase—by a factor of 2.7—between 1975 and 1986.[11] During

this same 1975-86 period, DoD funds for basic research—often referred to as 6.1 funding—in academia increased by a factor of 2.2 in constant dollars, while total research funds to academia—for basic plus applied research, or 6.1 plus 6.2 funding—increased by a factor of 2.1.[12] These DoD and NSF funding increases were not equally distributed in all fields, however, as Table 1 indicates; similar assertions hold for other federal agencies. As a matter of fact, in 1986 the total support of basic research in the physical sciences by all federal agencies, measured in constant dollars and including nonuniversity as well as university funding recipients, was no greater than it had been in 1967.[13]

COMPETITIVENESS AND THE ESSENTIALITY OF FUNDING SUPPORT

As a result of the outpourings of new, quite unrestricted research funds, university physical scientists enjoyed halcyon times during the years between the end of World War II and about 1965, when federal funds began to level off.[14] Con-

10. U.S. National Science Foundation, Division of Science Resources Studies, *Federal Funds for Research and Development, Detailed Historical Tables: Fiscal Years 1955-1988,* 1987, tab. B, pp. 22-38. The factors for converting from actual expenditures to constant dollars may be inferred from U.S. Department of Defense, *Report on the University Role in Defense Research and Development,* prepared for U.S. Congress, Committees on Appropriations, Apr. 1987, tab. 1, p. 7.

11. National Science Foundation, *Federal Funds for Research and Development.*

12. National Science Foundation, *Federal Funds for Research and Development,* p. 202, tab. 20A; p. 205, tab. 20B; p. 252, tab. 30A; p. 255, tab. 30B. See also Department of Defense, *University Role in Defense Research,* tab. 2, p. 10, and tab. 3, p. 14; John P. Holdren and F. Bailey Green, "Military Spending, the SDI, and Government Support of Research and Development: Effects on the Economy and the Health of American Science," *FAS Public Interest Report,* 39(7):1 (Sept. 1986).

13. National Science Foundation, *Federal Funds for Research and Development,* p. 272, tab. 35A, and p. 274, tab. 35B.

14. The period from about 1955 to about 1967 has been called "the golden age of academic science in the United States." Harvey Brooks, "The Dynamics of Funding, Enrollments, Curriculum and Employment," in *Physics Careers, Employment and Education,* ed. Martin L. Perl (New York: American Institute of Physics, 1978), p. 94.

comitantly, the existence of these new governmental funding sources initiated important still-persisting changes—from pre-World War II days—in the research performance patterns of universities and their quantitative science professors. Many of these changes have been examined, in much greater detail than is possible here, in studies sponsored during the 1970s by the Carnegie Commission on Higher Education[15] and the Alfred P. Sloan Foundation.[16] These changes also have been examined very recently by Forman,[17] in an interesting study that is much in the spirit of this article but with whose conclusions we do not wholly agree.

One major change has been a shift from a system in which the universities and private foundations were the major suppliers of university research funds in the quantitative sciences—these research funds being small fractions of the total departmental budget—to a system in which the major source of university research funds is the federal budget, and in which the funds supplied represent a substantial proportion of the total departmental budget. Indeed, in 1979 the federal government supplied 70.6 percent and 85.3 percent of the total R&D physical science expenditures at the leading public and private research universities, respectively.[18] These figures illustrate the fact

that—except for start-up moneys commonly awarded newly hired faculty—university physical science professors no longer expect, and no longer receive, significant research support from their home institutions' so-called hard-money funds.

Simultaneously, the way of life of most university quantitative science researchers also has greatly changed since pre-World War II days. As the grant system has evolved in this nation, academic principal investigators in the physical sciences are responsible for generating their own research funds, with which they purchase their equipment and support their graduate students, often even paying their students' tuition.[19] Most universities still offer teaching assistantships to some number of entering physical science graduate students, depending on teaching needs; to make room for the next batch of entering graduate students, however, the bulk of the already hired teaching assistants very soon must find thesis advisers who can support the students on their research grants, although a few graduate fellowships offering support on university hard money usually are available.

Physical science research is extremely competitive, moreover; priority in publishing is the name of the only game. The researcher who first publishes a new effect or a more accurate value of a physical constant will be acclaimed; his or her successors who merely repeat the previously published results, even if they reach those results more elegantly, at best will be footnoted and may not be published at all. Consequently, serious academic physical science experimenters

15. Dael Wolfle, *The Home of Science* (New York: McGraw-Hill, 1972).

16. Bruce L.R. Smith and Joseph J. Karlesky, eds., *The State of Academic Science,* vol. 1, *Summary of Major Findings* (New Rochelle, NY: Change Magazine Press, 1977); ibid., vol. 2, *Background Papers* (1978).

17. Forman, "Behind Quantum Electronics." See also the other papers in *Historical Studies in the Physical and Biological Sciences,* 1987, vol. 18, pt. 1, in which Forman's paper appeared.

18. Marilyn McCoy, Jack Krakower, and David Makowski, "Financing at the Leading 100 Research Universities," *Research in Higher Education,* 16(4): 340 (1982).

19. The evolution we describe was not inevitable. In Britain, for example, graduate students are not supported via charges to their thesis advisers' grants. Instead, a three-year grant is awarded directly to the graduate student, from funds supplied by the national government.

absolutely must have research funds; the days of doing experiments on a shoestring, with equipment painstakingly constructed in university machine and glass-blowing shops are gone forever, not merely because universities will not fully support such shops but also because investigators cannot take the time to construct their own apparatus whenever—as is now more and more often the case—the same or even better equipment is commercially available for purchase. Graduate students, who help design, set up, and run experiments, also are essential for expeditious accomplishment of experimental research; were they not, principal investigators hardly would be willing to devote their energies to finding funds for their students. In fact, because competent physical science graduate students have been in short supply for some time, principal investigators actually compete seriously for graduate students, especially for competent American-born students who need less training in laboratory work than foreign students and have no language difficulties.[20] The most common response by applicants for NSF support to a questionnaire inquiring about factors hindering "progress in your own research work" was "shortages of capable graduate students," listed by 47 percent of the respondents, and ahead of, for example, "inadequate equipment," listed by only 36 percent.[21]

Mathematics is similarly competitive; the Fields Medal[22] is awarded for the first proofs of famous conjectures, not for later—even if more instructive and useful—reformulations of those earlier proofs. It is true that until very recently mathematicians and physical science theorists—the latter being researchers like Einstein who develop new theories and/or predict their experimental consequences—had no need of equipment and could perform their research alone or with a very few disciples. Nowadays, however, most serious theoretical physicists and chemists also have need of significant funding support, because they cannot compete without recourse to the best computing facilities available; world-class-level computing can be very costly and can necessitate the hiring of considerable graduate-student help for programming, analyzing computer runs, and so forth. Even pure mathematicians, and perforce applied mathematicians, now are likely to make use of computers.[23] The NSF recently has awarded a $1.5 million three-year grant to the University of Minnesota for the so-called Geometry-Supercomputer Project, whose purpose is to use the power of a supercomputer to illuminate certain basic questions in

20. Timothy M. Beardsley, "Foreign-Born Students Fill U.S. Science Programs," *Scientific American,* May 1988, p. 22; Arthur M. Hauptman, "Students in Graduate and Professional Education: What We Know and Need to Know" (Report, Association of American Universities, 1986), p. 20; Elinor G. Barber and Robert P. Morgan, "Boon or Bane" (Report, Institute of International Education, 1988), pp. 18-21.

21. *Proposal Review at NSF: Perceptions of Principal Investigators,* NSF report 88-4, Feb. 1988, p. 28. Not all the quantitative sciences have

equal needs for graduate students, however. An excellent discussion of the differences between the disciplines in this regard is given by David W. Brenaman, "Effects of Recent Trends in Graduate Education on University Research Capability in Physics, Chemistry, and Mathematics," in *State of Academic Science,* ed. Smith and Karlesky, 2:133.

22. The mathematics community's analogue of the Nobel Prize, regarded as the most prestigious award in mathematics. Gina Kolata, "Mathematicians Recognize Major Discoveries," *Science,* 233:722 (1986); Edward E. David, Jr., "The Federal Support of Mathematics," *Scientific American,* 252(5):45 (May 1985).

23. Lynn Arthur Steen, "The Science of Patterns," *Science,* 29 Apr. 1988, p. 611.

geometry.[24] Pure mathematicians also are more and more frequently performing computer experiments to test their conjectures; as one mathematician who uses computers has observed, "'It's far easier to prove a difficult theorem if you know it's true.'"[25] It is the case, however, that for the most part pure mathematicians still rely less on graduate-student help and computers than do physical scientists; accordingly, modern pure mathematicians generally depend less on federal funds for the successful completion of their researches than do physical scientists.

The intense competition in the physical sciences and mathematics that has just been described and that always has existed has been considerably enhanced, especially in the physical sciences, by the post-World War II availability of federal funding. The large number of graduate students who had been supported in the early post-World War II years received their Ph.D.s and went on to do research on their own, either in newly created professorial positions wherein they could seek grants and train even more Ph.D. students, or else in industrial and government-supported laboratories, including FFRDCs, where they could do research full-time without having to worry about teaching responsibilities.

The resultant explosive increase in the number of physical scientists and the pace of research is illustrated by the history of the American Physical Society, the professional society to which most working U.S. physics researchers belong, founded in 1899. In 1940, it had 3751 members; by 1950, 1965, and 1987, this number had increased to 9472, 21,870, and 39,767, respectively.[26] The number of papers published in the *Physical Review,* the society's journal of research, increased roughly proportionally, growing from 525 in 1940 to 2749 and 7348 in the respective years 1965 and 1986.[27] The number of institutions granting Ph.D.s in physics increased from 28 in the early 1920s to 167 in the early 1970s. In 1920, these graduate schools awarded 31 Ph.D.s; in 1970, 1715 newly created physics Ph.D.s arrived on the scene.[28] In reality, the competition in the quantitative sciences during the last few decades has been even more intense than these numbers suggest, because of the growth in the number of quantitative scientists and publications in many foreign countries, for example, India and, of course, Japan.

24. Allyn Jackson, "Geometry-Supercomputer Project Inaugurated," *Notices of the American Mathematical Society,* 35(2):253 (Feb. 1988).

25. Stuart Geman, of Brown University, quoted in Allyn Jackson, "Supercomputers and the NSF," *Notices of the American Mathematical Society,* 34:9 (1987).

26. *Bulletin of the American Physical Society,* 32:2221 (1987).

27. *Physical Review,* 1940, vol. 57; ibid., vol. 58; *Bulletin of the American Physical Society,* 32:1329 (1987). These totals include the papers published in *Physical Review Letters,* a journal initiated by the American Physical Society in 1958 to enable rapid publication of important short communications, which previously had to appear in the less rapidly printed *Physical Review.* Actually, the growth in the number of physics papers published in this country has been considerably greater than even these numbers indicate, because the American Institute of Physics, an umbrella organization of physical societies to which the American Physical Society belongs, also has introduced new research journals since World War II, for example, the *Journal of Mathematical Physics,* begun in 1960, which in 1987 published 392 papers.

28. Similar increases occurred in chemistry and mathematics during the 1920-70 period. The number of institutions granting the Ph.D. increased from 43 to 194 for chemistry and from 22 to 159 for mathematics. The number of Ph.D.s awarded in chemistry increased from 77 to 2284 and in mathematics from 19 to 1282. U.S. National Academy of Sciences, *A Century of Doctorates: Data Analyses of Growth and Change,* 1978, tab. 2A, p. 12, and tab. 39, p. 95.

To a considerable extent, the new professorial positions referred to in the previous paragraph became available in university quantitative science departments because World War II's weaponry successes had sparked an explosion of interest in science and because the federal grant system permits universities to defray so much of the hard-money costs of supporting quantitative science research. Most of the federal granting agencies permit the university to charge some fraction of a principal investigator's salary to his or her grant. The grants also pay overhead, which helps to fund the university's central facilities. Overhead charges to grants are audited, of course, to ensure that they legitimately are assessable to the costs of grant administration; in actuality, however, many of the allowed overhead expenses, such as maintenance and heating costs in buildings where faculty offices and laboratories are located, would have to be defrayed by the university even if grants had not been received. Most universities now maintain administrative offices of research charged with the responsibility of facilitating faculty efforts to secure research funding.

In fact, a substantial fraction of the total budgets of the leading research universities now comes from federal sources, as indicated earlier. In 1979, the average proportions were 21.0 percent for the public institutions and 34.4 percent for the private, while for the particular private institutions of Cal Tech, Stanford, and the Massachusetts Institute of Technology (MIT), the fractions were 62.0 percent, 52.7 percent, and 51.2 percent, respectively.[29] It is not surprising, therefore, that many universities now balk at awarding tenure to quantitative science professors who do not have a good track record in obtaining grants. This criterion often is applied not only to physical scientists but also to pure mathematicians who have no great need of grants, on the not illogical—though inherently unfair—grounds that under the peer-review system the amount of research funding garnered by a principal investigator is a good measure of the quality of the investigator's research program.

MISSION AGENCY FUNDING INFLUENCES

Evidently, quantitative scientists in universities, especially physical scientists, continually are under great pressure to secure and/or maintain research support funds.[30] Beginning physical science graduate students have no possibility of securing research support funds on their own, however. Thus they quite naturally gravitate to those physical science professors who can support them; correspondingly, only those academic principal investigators who have grants are likely to be producing students with advanced physical science degrees. Furthermore, the owners of those new advanced degrees, who now have a hard-earned investment in the same specialty as their former thesis advisers, are likely to continue working in that specialty after they leave the university. It follows that the funding criteria of the various federal agencies that support university science research, which criteria depend on the individual agency's mission, can greatly influence the directions of research in an entire broad field of physical science, such as chemistry or astronomy.

29. McCoy, Krakower, and Makowski, "Financing at the Leading 100 Research Universities," pp. 334-35.

30. An established physicist who changed fields has told one of us (EB), "If it hadn't been for the never-ending proposal-writing rat race, I never would have left physics to become a lawyer."

The mission of the NSF is spelled out in its original charter: "to develop and encourage the pursuit of a national policy for the promotion of basic research and education in the sciences."[31] Deciding what research the NSF will fund is a complex process. It involves the setting of policy by the National Science Board, many of whose members are eminent scientists, followed by detailed assessments of need by program managers, often active scientists themselves. The program managers base their recommendations on ideas received from professional societies, discussions by program panels, the comments of proposal referees, and the program managers' own contacts with competent scientists.[32] While the NSF budget must win congressional approval, which may result in a bias toward applied research because the need for basic research is harder for laypersons to grasp, the planning of what areas the NSF will fund is influenced by working scientists at all stages of the planning process. In other words, because of its mission and its selection procedures, NSF funding priorities are to a very great degree directed toward supporting the best and most exciting research, as determined by the scientific community itself.

In contrast to the NSF, the missions of other federal agencies are not primarily directed toward the promotion of basic scientific research; these other agencies fund basic science only when such funding advances these agencies' actual primary missions, for example, in NASA's case, only when the research advances NASA objectives like "'the development and operation of vehicles capable of carrying instruments, equipment, supplies, and

living organisms through space.'"[33] The primary mission of the DoD and its funding agencies is the maintenance of national security, which need not be directly related to scientific research. In furtherance of this mission, however, these DoD agencies do support basic research related to long-term national security requirements. For example, ONR has the statutory responsibility "to plan, foster, and encourage scientific research in recognition of its paramount importance as related to the maintenance of future naval power and the preservation of national security."[34] As discussed earlier, in the period immediately following World War II, this directive was broadly interpreted by ONR, as were the similar directives to the Army Research Office and the Air Force Office of Scientific Research. Thus, in the early years after World War II, the funding priorities of these DoD agencies were not distinguishable from those developed by the NSF. As time progressed, however, and particularly after the 1970 Mansfield Amendment's requirement of a "direct and apparent relationship" between DoD-funded research and DoD missions, relevance to possible military applications became an essential criterion for research support by the DoD funding agencies.

After the Mansfield Amendment, "it was no longer sufficient that the programs' relation to the present or future Navy be clear to the scientists in the Navy and Defense Department; the relation had to be reasonably clear to those who chanced to look."[35] Thus, although the "merit review"[36] process does provide for

31. National Science Board, *Basic Research in the Mission Agencies,* p. 191.
32. Ibid., p. 268.

33. Ibid., p. 184, quoting the National Aeronautics and Space Act of 1958, 42 U.S.C. §2452.
34. Ibid., p. 267.
35. Ibid., p. 286.
36. U.S. Department of Defense, *The Department of Defense Report on the Merit Review*

peer evaluation of the scientific merit of proposals, under the present *modus operandi* of the DoD funding agencies, the areas to be funded are influenced by the agencies' national security mission.[37] In fact, the DoD itself has explained that "an external peer review process as the basis for awarding university research funds would be appropriate" if an agency's mission were only "to support good science," whereas DoD's merit review process also properly takes into account the relevance of the proposed research to DoD's mission.[38]

Extent of dependence on DoD

How greatly the mission orientation of the DoD funding agencies skews university quantitative science research depends on how heavily such research relies on DoD funding. Table 1 lists, in constant dollars, the DoD funds expended for research performed at universities and colleges during 1976 and 1986 in each of the quantitative science fields, namely, physics, chemistry, astronomy, and mathematics.[39] For comparison, we give figures also for computer science, electrical engineering, mechanical engineering, metallurgy and materials engineering, and the environmental sciences, in which are included primarily the atmospheric sciences, geology, and oceanography. We have followed the lead of Yudken and Simons in including the NSF category "Mathematics and Computer Science, Not Elsewhere Classified" in "Computer Science."[40] Also shown in Table 1 are the percentages of DoD and NSF funds in the total support figures for research performed at universities in fiscal year 1986, as well as the percentages of all federally funded full-time graduate students in doctorate-granting institutions who are funded by the DoD and the NSF.[41]

We now see that the dramatic increase in DoD funding of university research in the decade preceding 1986, discussed earlier, differs for individual fields in the quantitative sciences. Similarly, the fields show a different dependence on DoD funding; for some, the DoD is a major contributor of federal funds, while for other fields DoD funds are a relatively minor portion of the total support. It is noteworthy—in fact, surprising to us in view of the World War II and post-World War II influence of physicists on weapons R&D[42]—that DoD funding in constant

Process for Competitive Selection of University Research Projects and an Analysis of the Potential of Expanding the Geographical Distribution of Research, prepared for U.S. Congress, Committees on Appropriations, Apr. 1987.

37. For an instructive, detailed description of how the military agencies select the research proposals they fund, see Stanton A. Glantz and Norm V. Albers, "Department of Defense R&D in the University," *Science,* 186:706 (1974).

38. Department of Defense, *Report on the Merit Review Process.* It has been argued, however, that relevance lies in the eye of the beholder and that therefore the DoD's merit review process need not lead to different funding patterns from those led to by the NSF's peer review. John Walsh, "Pentagon Plans Boost for Basic Research," *Science,* 205:566 (1979).

39. U.S. National Science Foundation, Division of Science Resources Studies, *Federal Funds for Research and Development: Federal Obligations for Research to Universities and Colleges by Agency and Detailed Field of Science/Engineering: Fiscal Years 1973-1987,* n.d., tab. 1A; *Federal Funds for Research and Development: Fiscal Years 1986, 1987 and 1988,* NSF Science Resources Series, 1987, tabs. C-86, C-89, and C-92.

40. Joel S. Yudken and Barbara B. Simons, "Federal Funding in Computer Science: A Preliminary Report," *SIGACT News,* 19(1):53 (Fall 1987).

41. *Academic Science/Engineering: Graduate Enrollment and Support,* NSF Science Resources Series, 1988, tab. C-18. Unfortunately, this report gives data for broad fields only.

42. For example, a large fraction of the directors of the Los Alamos and Lawrence Livermore weap-

TABLE 1
ANALYSIS OF DoD SUPPORT FOR UNIVERSITY RESEARCH

	DoD Support for University Research (Millions of 1987 dollars)		Percentage Share of Federal Support for University Research, Fiscal Year 1986		Percentage Share of Federally Supported Graduate Students, Fall 1986	
	Fiscal Year 1976	Fiscal Year 1986	DoD	NSF	DoD	NSF
Physical sciences*	60.0	101.0	13.0%	37.9%	12.0%	38.8%
Astronomy	0.6	6.3	9.0	39.6		
Chemistry	15.3	44.6	16.3	38.9		
Physics	41.0	49.2	11.8	37.8		
Mathematics	18.6	36.2	33.3	48.6	43.0	36.0
Computer science	23.7	118.2	67.8	22.3	55.8	30.0
Engineering†	112.8	284.2	45.0	28.7	33.1	25.2
Electrical engineering	37.4	93.9	65.6	28.6		
Mechanical engineering	15.6	48.1	56.3	27.5		
Metallurgy and materials engineering	25.5	66.4	48.3	28.9		
Environmental sciences	61.7	92.4	24.5	55.2	15.0	44.8
Total (all fields)‡	369.2	730.7	12.6	17.7	15.3	21.0

SOURCES: For support for university research: U.S. National Science Foundation, Division of Science Resources Studies, *Federal Funds for Research and Development: Federal Obligations for Research to Universities and Colleges by Agency and Detailed Field of Science/Engineering: Fiscal Years 1973-1987*, n.d.; *Federal Funds for Research and Development: Fiscal Years 1986, 1987 and 1988*, NSF Science Resources Series, 1987. For percentage share of federally supported graduate students: *Academic Science/Engineering: Graduate Enrollment and Support*, NSF Science Series, 1988.

*Includes fields "not elsewhere classified" in addition to listed fields.
†Includes all engineering fields.
‡Includes all fields funded, in addition to listed fields.

dollars has increased least rapidly for physics, and that—excepting astronomy, on which the DoD spends comparatively little—DoD funds also are least important for physics. The contrasting DoD funding

ons laboratories have been physicists. Harold Brown, President Carter's secretary of defense, has a physics Ph.D. In addition, the Jason group, one of the DoD's most important military science advisory committees, is composed of about fifty of this nation's most elite physicists, almost all from academia. Paul Selvin and Charles Schwartz, "Publish and Perish," in *Military Funding of Scientific Research*, special issue of *Science for the People*, 20(1):10 (Jan.-Feb. 1988); Kevles, *Physicists*, p. 402.

increases and support shares for physics and for computer science merit notice as well. In the quantitative sciences, DoD funding has had its largest percentage increase during the last decade in chemistry—again excepting astronomy—but presently makes the largest fractional contribution to total federal funding in mathematics. It also is noteworthy that for the various fields listed in Table 1 there is a close correspondence—which for the physical sciences even is a near equality—between the percentages of all full-time graduate students supported on federal funds who are supported by the DoD and the percentages of all federal university research funding that are attributable to the DoD. This correspondence confirms our earlier inference, based on our understanding of graduate-student psychology and needs, that DoD funding of fields or subfields draws graduate students to those research areas.

Of course, because the DoD does not fund equally in all subfields of a scientific discipline, the small percentages of DoD support in physics and in chemistry shown in Table 1 are deceiving to some extent. For instance, the large well-funded area of elementary-particle physics is not supported by the DoD, which concentrates on atomic physics and condensed-matter physics.[43] In these latter areas of physics, the percentage from the DoD of all federal support for university research appears to be 30 percent, rather than the 11.8 percent figure of Table 1 for physics as a whole.[44] Furthermore, the numbers that have just been quoted include only research and exclude development; consequently, these figures are an underestimate of total DoD support to universities, especially in engineering. In addition, the

figures do not include Strategic Defense Initiative funding, which is labeled "advanced development," or 6.3a, funding. All university research funded via the Strategic Defense Initiative is included in the so-called Innovative Science and Technology program, of which between about 67 percent and 80 percent is awarded to universities.[45]

The magnitude of DoD support varies greatly not only between the disciplines but also between universities. In fiscal year 1986, Columbia, Cornell, and Yale obtained only a minor proportion—less than 10 percent—of their federal funds for R&D from the DoD; MIT, Stanford, and Cal Tech received a considerably larger proportion—20-25 percent—while some other major institutions—notably Carnegie-Mellon, Johns Hopkins, and Georgia Tech—received more than 50 percent of their funds from the DoD: 61 percent, 71 percent, and 73 percent, respectively.[46] A number of institutions that are less prestigious than those just named also rely very heavily on DoD R&D support, for example, the University of Dayton, for 94 percent of its funds; New Mexico State University, for 67 percent; and Utah State University for 57 percent. The funds for university-administered FFRDCs, many of which are heavily

43. U.S. Air Force Office of Scientific Research, *Research Interests of the Air Force Office of Scientific Research,* pamphlet 70-1, Jan. 1988; U.S. Army Research Office, *United States Army Broad Agency Announcement,* July 1987; U.S. Department of Defense, *Defense University Research Instrumentation Program, Fiscal Year 1988,* announcement available from ONR Code 11SP, 1988.

44. Using published breakdowns of the research budgets of various agencies by area within physics, we estimate that approximately 30 percent of the DoE research funds, 30 percent of the NSF funds, and 100 percent of the DoD and NASA funds awarded to physics researchers in universities fall within the fields of atomic and condensed-matter physics. Irwin Goodwin, "Washington Reports," *Physics Today,* 41(4) (Apr. 1988). Using these just-quoted figures and the tables listed in fn. 39 of this article, we arrive at the 30 percent figure quoted in the text.

45. Holdren and Green, "Military Spending"; Robert Park, "Physics in the FY 1988 Budget," in *AAAS Report XII, Research and Development FY 1988* (Washington, DC: American Association for the Advancement of Science, 1987), p. 157.

46. *Federal Support to Universities, Colleges, and Selected Nonprofit Institutions, Fiscal Year: 1986,* NSF Surveys of Science Resources Series, Jan. 1988, tàb. B-9. Much of the DoD funding received by Johns Hopkins is expended in its Applied Physics Laboratory, however, which is not classified as an FFRDC, though its operations and functions are very similar to most FFRDCs; if the Applied Physics Laboratory funding is subtracted, the 71 percent figure stated in the text would be significantly reduced.

supported by the DoD, are not included in the just-quoted percentages. For some universities, including FFRDC funding would greatly increase the percentages of DoD funds in the total amount of federal support; or, in the case of FFRDCs like the Los Alamos and Lawrence Livermore Laboratories that, though funded through the DoE, are heavily engaged in weapons R&D, including FFRDC funding would greatly increase the percentages of DoD-equivalent DoE funds.[47]

The foregoing actual and potential DoD influences on universities and their quantitative science departments are amplified, moreover—especially in physics—by recent trends that are not well measured merely by the amounts and percentages of DoD funding. Many of the FFRDCs that are intimately involved in weapons R&D now enjoy such very advanced equipment and computing facilities, and employ so very many highly skilled researchers, that in some quantitative science fields university researchers, particularly those in smaller, less prestigious universities, are unable to compete without a formal association with those FFRDCs via, for example, consulting contracts or summer employment.[48]

Each of the services supports a number of well-paying, competitively awarded predoctoral and postdoctoral fellowships, which are limited, however, to research areas of interest to the DoD.[49] The Los Alamos National Laboratory offers similar fellowships;[50] the University of California at Davis allows physics and engineering students to conduct their Ph.D. work at the Lawrence Livermore Laboratory. The prestigious, well-paying Hertz graduate fellowships, although funded by private moneys, are administered by Livermore Laboratory with the stated purpose of improving "the defense potential of the United States."[51] Since 1983, the DoD has been awarding approximately $25 million a year for upgrades of university research facilities in areas of interest to the military;[52] such sums for renewal of research facilities, which in many universities have become quite outmoded, were more than comparable to the research

---

told one of us (EG), "Not only did I significantly increase my salary, but I now have over thirty atomic physics colleagues I can talk to. At the university I was the only one in my field."

49. For example, the ONR Young Investigator Program and the Air Force Laboratory Graduate Fellowship Program, described in Robert Krinsky, "Swords into Sheepskins," in *Science for the People,* 20(1):5 (Jan.-Feb. 1988). See also, for example, Air Force Office of Scientific Research, *Research Interests of the Air Force Office of Scientific Research.*

50. Cf., for example, the advertisements in *Physics Today,* Jan. 1988, p. 151, and ibid., Apr. 1988, p. 141, as well as in *Science,* 22 Apr. 1988, p. 565. These advertised postdoctoral fellowships carry a salary of $32,820 to $34,580 per annum; exceptionally qualified candidates may receive J. Robert Oppenheimer fellowships at an annual salary of $50,600.

51. Gary Marchant, "Political Constraints," in *Science for the People,* 20(1):32 (Jan.-Feb. 1988); *Annual Register of Grant Support, 1987-88* (Wilmette, IL: National Register, 1988), p. 646.

52. Department of Defense, *Defense University Research Instrumentation Program,* p. 4.

47. According to a recent report from the House of Representatives' Energy and Power subcommittee, in the Reagan administration's proposed fiscal year 1989 DoE budget, $8.2 billion, or 61 percent of the total DoE request, represented activities in support of the DoD. *Bulletin of the American Physical Society,* 33(5):1156 (May 1988). In 1985, DoD funding of Livermore projects amounted to $192 million, which equaled 29 percent of DoE's own funding of Livermore; the corresponding 1985 figures for Los Alamos were $178 million and 33 percent. General Accounting Office, *Research and Development Centers,* pp. 35-36.

48. An internationally recognized atomic physicist, who recently has resigned from his professorship at a medium-prestige university in favor of an appointment at the Lawrence Livermore Laboratory after having been a consultant for a time, has

renewal funds available from the NSF.[53] The fact that the DoD can attract such highly prestigious scientific advisers, as, for example, the physicists constituting the Jason group[54] enhances the scientific importance of research in areas of DoD interest.

The foregoing analysis of the numbers listed in Table 1, taken together with our discussions under previous subheadings of this article and considerations such as those in the preceding paragraph, leads us to the measured conclusion that— except probably in astronomy—support by the mission-oriented DoD agency is significantly but by no means overwhelmingly skewing research directions and subfield growths in all the quantitative sciences. We correspondingly conclude that DoD funding probably is having some influence on the recruiting and promotion policies of many university quantitative science departments and, especially in those universities that rely very heavily on DoD support, may be influencing top-level administrator recruiting and administrative policies as well. Our conclusion about the skewing effects of DoD support on the quantitative sciences is more moderate than seemingly would be warranted in some other fields, for example, computer science.

SPECIAL PROBLEMS OF DoD FUNDING

Many thoughtful quantitative scientists, though they may not have reached precisely the same conclusions as we, are troubled by the implications of DoD research support. In January 1988 the American Mathematical Society undertook a mail referendum to its 21,400 members on, *inter alia,* the following statement:

The AMS is concerned about the large proportion of military funding of mathematics research. There is a tendency to distribute this support through narrowly focused (mission oriented) programs and to circumvent peer review procedures. This situation may skew and ultimately injure mathematics in the United States. Therefore those representing the AMS are requested to direct their efforts toward increasing the fraction of non-military funding for mathematics research, as well as toward increasing total research support.[55]

The vote on this statement was 5193 for and 1317 against.[56] Although academic spokespersons continue to press for increases in DoD funding of university research,[57] this lopsided vote—and other actions by university quantitative scientists that will be described herein—indicate that concerns about DoD funding are widely held in the community of working quantitative science researchers. Of particular significance is the fact that to our knowledge such concerns have not been expressed about funding by any other federal mission agency, such as NASA, although—for reasons already explained—support by any mission agency is capable of skewing research directions and affecting university policies.

53. "Financing and Managing University Research Equipment" (Report, Association of American Universities, 1985), pp. 144-64; *AAAS Report XIII: Research and Development, FY 1989* (Washington, DC: American Association for the Advancement of Science, 1988), pp. 222-23.

54. See fn. 42 of this article.

55. *Notices of the American Mathematical Society,* 34(6):1014 (Oct. 1987).

56. *Chronicle of Higher Education,* 6 Apr. 1988, p. A4.

57. Robert M. Rosenzweig, "Statement on Behalf of AAU, AGE, NASULGC, AGS and ACS," testimony presented to U.S. Congress, House, Committee on Armed Services, Subcommittee on Research and Development, 17 Mar. 1988; "Rosenzweig Testifies before HASC R&D Subcommittee; Subcommittee Marks up FY89 Authorization," *AAU Washington Report,* 25 Mar. 1988.

Restrictions on information transfer

It is likely that the special concerns about DoD funding stem from some substantial special problems associated with military support of university research. One such special problem, posed particularly by DoD funding of university quantitative science research, is the possibility that such support may involve restrictions on the performers of the funded research as well as on the dissemination of the results obtained. As the technology produced by science has become increasingly important to the national defense, the national welfare increasingly has required that some technical information be kept from our democracy's enemies, whether foreign or domestic; certainly, it would be difficult to find Americans who believe that secessionist fringe sects or the Ayatollah Khomeini should have access to the secrets of constructing hydrogen bombs. Until quite recently the government's primary methods for preventing the uncontrolled transfer of information that "reasonably could be expected to cause damage to national security"[58] was through classification, as is elaborated in another article in this *Annals* issue; on the grounds of national security certain information is unmistakably classified secret, to be revealed only to persons with security clearance and with a need to know.

The mission of our research universities is the education of the future generation of scholars and the advancement of knowledge. The restrictions that secrecy classifications impose on the publication of research results and the exchanges of information with students and colleagues, whatever their national security justifica-

tion, are quite incompatible with this mission. Because classified information is so readily identifiable, however, universities have been able to handle classification requirements without much difficulty. Many universities have taken the simplest and most obvious step of totally banning classified research from their campuses;[59] even if not banned, the volume of classified research on all but a few campuses must be very small.[60]

There are statutes and regulations, however, most notably those controlling U.S. exports, that restrict the dissemination of unclassified technical information; for the purposes of this article, it is convenient to call such controlled unclassified information "sensitive," although in actual technical-data-transfer enforcement practice the term "sensitive" has a precise, more limited definition. Illustrative governmental applications of sensitive-information controls to restrict the research activities of university quantitative scientists are described elsewhere in this *Annals* issue. Here it suffices to say that universities have found restrictions on sensitive information more difficult to deal with than restrictions on classified information.

The difficulties stem from the circumstance that the applicability of, for example, export-control requirements often becomes apparent—if apparent at all—

58. President Reagan's Executive Order no. 12,356, §1.3(b), 47 Fed. Reg. 14.874 (1982).

59. "Report of the Committee on Threats to Autonomy in University Research" (Association of Graduate Schools, 1987).

60. Mitchel B. Wallerstein and Lawrence E. McCray, "Scientific Communication and National Security: The Issues in 1984" (Report, Committee on Science, Engineering and Public Policy, National Research Council, Jan. 1984), p. 14. This analysis of 123,000 DoD reports showed that of the 23,119 originating in universities only 1.3 percent were classified; every classified report was generated in off-campus facilities affiliated with universities—for example, FFRDCs—not on university campuses as such.

only after the research has been completed and thus cannot be avoided in advance. This circumstance stems in turn from the fact that sensitive information, unlike classified information, is not readily identifiable, because the regulations delineating sensitive from nonsensitive information are extremely difficult to disentangle. Moreover, although the restrictions on dissemination of sensitive data can apply to any quantitative science researcher, those with DoD funding do seem to be particularly vulnerable; because of the DoD's stated mission, taken together with the relevancy requirement originally stemming from the Mansfield Amendment, DoD-funded researchers are more likely than others to be doing research that may be deemed sensitive.[61] As a matter of fact, in the recent past the DoD, agreeing with his assessment, has seriously discussed requiring many of its university principal investigators to accept prepublication review of research results as a condition for receipt of even unclassified research contracts.[62] Especially disturbing to the entire academic community, not just to quantitative scientists, has been the fear—largely fostered by the DoD's own unnecessarily abrasive statements[63]—that funded principal investigators and universities ultimately would become beholden to and dependent on the DoD.

61. "National Security Controls and University Research: Information For Investigators and Administrators" (Report, Association of American Universities, June 1987), p. 8.

62. See the discussion of attempts at information control in David Wilson, "Consequential Controversies," this issue of *The Annals* of the American Academy of Political and Social Science.

63. See, for example, the 1986 remarks of then Under Secretary of Defense Donald Hicks, quoted in Vera Kistiakowsky, "Military Funding of University Research," this issue of *The Annals* of the American Academy of Political and Social Science.

Campus opposition to DoD funding

A second—and very important—problem that DoD funding of research poses for universities stems from the fact that in most universities acceptance of such support is likely to be opposed by segments of the faculty and student body. Although these segments usually are not large, their opposition is frequently strident, always divisive, and not uncommonly accompanied by physically violent acts, including, for example, unlawful occupation of campus buildings, as in the well-publicized 1968 student uprising at Columbia University, which began with student protests against, *inter alia,* Columbia's involvement in defense research.[64] The restrictions on information transfer certainly are an element in this opposition to DoD research funding, but the more basic reason for the opposition undoubtedly is the discordance between what most faculty and students believe should be the objective of university research, namely, advancement of fundamental knowledge for the benefit of all humankind, and the DoD's main objective—protection of this nation's welfare at all costs, including, if necessary, the destruction of other nations.

The opposition to DoD support tends to be strongest and most vocal at times when the United States is involved in not wholly popular foreign policies or military engagements. At such times, moreover, the opposition often focuses on DoD-funded quantitative science research, doubtless because of the demonstrated contributions to weapons development

64. Morris Dickstein, "Columbia Recovered," *New York Times Magazine,* 15 May 1988, p. 32. There also were demonstrations against Columbia's faculty involvement with the Jason group (recall fn. 42). Kevles, *Physicists,* p. 403.

by quantitative scientists. In particular, during the Vietnam war the FFRDCs that were located on or near university campuses were major targets of student demonstrations. At the University of Wisconsin, such demonstrations culminated in 1970 with the death of a postdoctoral student in a bombing of the Army Mathematics Research Center located on campus,[65] after which the university phased out its administration of the center.

The University of Wisconsin violence was uniquely tragic. Nevertheless, such protests—together with the realization that the university's aforementioned research objective may not be consistent with close involvement of regular faculty and graduate students in the classified research that DoD often conducts—have caused many other universities to cut loose from the FFRDCs they had been administering. For example, in 1970 Stanford University severed its ties with the Stanford Research Institute, located a short distance from campus;[66] the MIT Instrumentation Laboratory, adjacent to the campus, was separated from the university in 1973 and was renamed the Charles Stark Draper Laboratory.[67] In fact, of the eight university-managed DoD-sponsored FFRDCs extant in 1966, five had been phased out or transferred to other management by 1975.[68] It was during this same period, also in response to student-faculty protests, that many universities took the previously mentioned step of banning or severely restricting on-campus classified research.

More recently, with the increase in the military budget and the dissatisfaction of many faculty and students with various aspects of the Reagan administration's foreign policy, there has been a resurgence of campus activity against DoD research funding, although not with as much intensity or with as wide a participation—especially among students—as during the Vietnam war. There has been especially strong opposition to funding connected with the Strategic Defense Initiative, or Star Wars, as is explained elsewhere in this issue. At a number of prestigious universities, the recent protests against DoD funding of research have coincided with faculty reexaminations of research support policies. Vera Kistiakowsky's companion article to ours discusses several such reexaminations. We content ourselves here with a description of the emotional campus debates and tense faculty meetings touched off at Cal Tech in the winter of 1983-84.

The events grew out of the early 1980s decision by the administration of the Jet Propulsion Laboratory (JPL)—a NASA-funded FFRDC located close to Cal Tech and managed by it—to compensate for decreases in NASA funding by increasing substantially the funding from military sources. Reluctant permission was given by the Cal Tech faculty to increase military funding to 30 percent of the total of JPL funding.[69] About a year later, the Army invited Cal Tech to administer a new Army think tank, similar to the long-established Air Force's Rand Corporation. In response to this invitation, the Cal Tech administration approved the establishment of the Arroyo

65. Kevles, *Physicists,* p. 403.

66. Walter S. Baer, "The Changing Relationship: Universities and Other R&D Performers," in *State of Academic Science,* ed. Smith and Karlesky, 2:78.

67. Ibid.

68. Ibid., p. 79.

69. John F. Benton, "Chronology of Army Analysis Program at Caltech" (Memo to voting faculty, California Institute of Technology, 27 Jan. 1984).

Center at JPL,[70] whose research projects largely would be classified and almost exclusively would involve practical military problems.[71]

A small but significant group of Cal Tech faculty members were outraged that this action, which so substantially changed the role and mission of JPL, was undertaken with almost no consultation with the faculty.[72] A memorandum to the Cal Tech faculty from its Graduate Student Council raised a number of objections to the Arroyo Center. Among them, for example, were that "many Graduate Students are concerned that much of military research is leading the world into extreme danger" and that "many Graduate Students feel that Cal Tech should not be developing a closer affiliation with a secret military research institute, and thus lending its prestige and name to such a facility."[73] Perhaps because of this campus reaction, Cal Tech's association with the Arroyo Center was terminated soon thereafter; management of the center was transferred to the Rand Corporation in fiscal year 1985.

It would be unwarranted to regard this history of Cal Tech's short-lived involvement with the Arroyo Center as an illustration of our conclusion that univer-

sity reliance on DoD funding may influence top-level administrative decision making. It would not be unwarranted to regard this history as an illustration of the fact that a university decision to increase DoD funding at an FFRDC it manages can spark student dissension and faculty reexamination of DoD research support policies on the main campus.

Not all the faculties of prestigious universities have been reexamining their institutions' research support policies, however, even when the occasions for such reexamination have been unusually opportune. In particular, Carnegie-Mellon University (CMU)—which already receives almost two-thirds of its federal research funding from the DoD and, except for Johns Hopkins and MIT, receives more DoD funds than any other university[74]—recently has been successful in a strongly pursued effort to be designated the administrator of a new FFRDC, the Software Engineering Institute (SEI), funded by the Air Force.[75] The institute's just-completed building is adjacent to campus, directly across the street from the building housing the biological sciences. The only remaining university-managed DoD-funded FFRDC is MIT's Lincoln Laboratory, which is situated more than ten miles from the MIT campus.[76] Many SEI projects are of a direct military nature, such as research that would reduce pilot error in targeting or allow a missile to track its target with more precision, and classified work will be permitted.[77]

70. Ibid.

71. M. L. Goldberger and R. E. Vogt, "Background Material for Faculty Discussion"(Memo to the faculty, California Institute of Technology, 23 Jan. 1984), esp. app. A, "Arroyo Center FY 1984 Program Content."

72. Lee Dembart, "Army Think Tank Plan Stirs Caltech," *Los Angeles Times,* 30 Jan. 1984.

73. Graduate Student Council, "Army Analysis Program, a.k.a. Arroyo Center" (Memo to Cal Tech faculty, n.d. but apparently written shortly after 12 Jan. 1984). The faculty and graduate student outcry may have been exacerbated by the fact that Cal Tech's agreement to administer the Arroyo Center closely coincided with the appointment—as vice-president of Cal Tech and director of JPL—of Lew Allen, Jr., former Air Force general and chief of staff, who had just retired.

74. Colleen Cordes, "Pentagon Awards $1.4 Billion in R&D Contracts," *Chronicle of Higher Education,* 13 Apr. 1988, p. A32.

75. The SEI was established as a new FFRDC, to be managed by CMU, in December 1984. General Accounting Office, *Research and Development Centers,* p. 42.

76. Ibid., p. 14.

77. Tela C. Zasloff, "The University, by Definition, May Be the Wrong Place for Military

There have been community objections to the SEI, manifested in picketing during construction of the SEI building,[78] but the issue of CMU's accepting administration of the SEI does not appear to have been thoroughly examined by the faculty. There has been a long and painful debate in the CMU Senate that terminated in faculty approval of a policy limiting to "semi-autonomous units" like the SEI any research, whether supported by industry or government, whose results cannot be freely disseminated.[79] It also has been reported[80] that some CMU faculty have been greatly embarrassed by a *New York Times* front-page story that—after stating that Martin Marietta had been selected as a prime contractor for Star Wars simulations—included the paragraph, "As the prime contractor, Martin Marietta will perform about half the work itself, and it will oversee a small army of subcontractors that include the Hughes Aircraft Company, the International Business Machines Corporation, the Carnegie-Mellon University and a host of smaller high-technology companies."[81]

The SEI's declared scope of effort, which in part is to "assess the potential of software technology that could aid the development and evaluation of mission-critical software,"[82] presumably is the reason CMU was chosen as Martin Marietta's "subcontractor." If history instructs

us, the SEI is likely to cause great trouble for CMU in the future, especially if the nation once again embarks on a not fully popular military venture, since the SEI's mission is so greatly at variance with CMU's mission as a university.

CONCLUDING OBSERVATIONS

In the years since World War II, American basic scientific research has attained world preeminence. In the quantitative sciences, with which we are concerned here, U.S. basic research journals are the most prestigious, publish the most papers, and are widely cited.[83] Since 1955, U.S. scientists have garnered half of all the Nobel prizes awarded in physics and chemistry, which are the only quantitative sciences eligible for a Nobel prize. Most of the published quantitative science basic research by U.S. scientists, as well as most of the research that has earned American physicists and chemists their Nobel prizes, has been performed in this nation's universities. Indeed, the widely accepted practice that has sustained the previously described post-World War II growth of U.S. university quantitative science departments—namely, that tenure cannot be earned without proficiency in research—has earned American graduate education in the quantitative sciences the respect and envy of all nations. Evidence for this assertion is the fact that science graduates from colleges all around the world flock to the United States for postgraduate training. In 1985, as many

Research," *Chronicle of Higher Education,* 20 Apr. 1988, p. A52.

78. Ibid.; the picketing has been seen by the authors of this article.

79. Carnegie-Mellon University, "Faculty Senate Minutes," 21 Jan. 1988; ibid., 29 Feb. 1988.

80. Private communication to one of us (EB) from a CMU faculty member.

81. David E. Sanger, "Martin Marietta Selected by U.S. for Tests Simulating 'Star Wars,'" *New York Times,* 23 Jan. 1988, p. 1.

82. General Accounting Office, *Research and Development Centers,* p. 28.

83. In 1982, U.S. scientists authored 27 percent of the entire world's production of physics papers. The corresponding figures for chemistry and mathematics were 21 percent and 37 percent, respectively. Physics papers published in U.S. journals during 1982 were cited 1.5 times as often as the average published paper in the field; the corresponding ratios for chemistry and mathematics were 1.63 and 1.23. *International Science and Technology Data Update 1987,* NSF report 87-319, 1987.

as 66,100 foreign science and engineering graduate students were enrolled full-time at doctorate-granting institutions in the United States; this number represents an increase of 8 percent from 1984, although the corresponding number of U.S. graduate students decreased 1 percent from 1984 to 1985.[84]

In short, our research university system is one of this nation's important strengths, as the DoD's active support of university quantitative science research attests. Unfortunately, this very support, though it does increase the amount of desperately needed funds available to researchers, simultaneously is affecting university quantitative science in several significant ways, as we have discussed: research directions are being skewed, departmental hiring and promotion policies probably are being influenced, and top-level administrative policies and recruiting may be influenced as well. It is not necessarily the case that these influences always or even frequently are adverse, but we risk the continued health of our heretofore highly successful research universities when we allow their policies to be swayed by extraneous factors stemming from the DoD's attempts to implement its mission, which is so different from the universities' mission of educating future scholars and advancing knowledge for the benefit of all humankind. Directly subjecting our research universities to such extraneous DoD influences, which also may affect campus morale and hinder free dissemination of research results, is wrong in principle and all too often may be deleterious in practice.

On the other hand, it is unlikely that this nation's political leadership, which now permits the DoD to spend so large a fraction of the federal budget, would agree that our research universities should be fully isolated from the DoD. The DoD believes that its isolation from our research universities would be unwise, as we have emphasized, and—once the necessity of continually modernizing our military forces is granted, as our citizenry apparently does grant—this belief of the DoD's is defensible, whether or not it ultimately is in the nation's best interests. Certainly, permitting the DoD to perform militarily related research solely on its own, free of oversight by probably the only civilian segment of the national community that can fully grasp the implications of those researches, could pose grave dangers to the nation.[85] The concluding task of this article, therefore, is to examine how the present national decision, probably unalterable for the immediate future, that our universities should not be totally isolated from military research can be reconciled with the obvious national needs to sustain the health of our entire research university system in general and of our currently internationally acclaimed quantitative science research departments in particular.

The relevancy requirement

A point of departure for this examination is the present requirement that individual basic research supported by the DoD must be relevant to the DoD's mission. At least insofar as the quantitative sciences are concerned, this relevancy requirement, however well intentioned, has been a troublesome mistake. The requirement, dating from the Mansfield Amendment, has not prevented

84. *Foreign Students Fueled 2% Rise in 1985 Graduate Science and Engineering Enrollment,* NSF Science Resource Studies Highlights, NSF 87-306, 12 June 1987, p. 2.

85. This very point was made during the early discussions about the pros and cons of the Mansfield Amendment. Rodney W. Nichols, "Mission-Oriented R&D," *Science,* 2 Apr. 1971, p. 31.

"undue dependency of American science on military appropriations," probably Mansfield's primary objective in espousing the amendment,[86] and there is no evidence that it has helped the DoD to spend its basic research support funds more effectively. Instead, the relevancy requirement merely has contributed to the skewing of quantitative science research directions, as we have described, even though—if World War II history is any guide at all—there is absolutely no basis for believing that the DoD can better predict what basic research directions will best advance those scientific and technical strengths of the United States upon which the DoD hopes to rely than can working American scientists, who are fully cognizant that this nation's strengths need to be maintained.

A simple solution to the dilemmas posed by the relevancy requirement might be to broaden the DoD's mission. For instance, the DoD—under congressional authority if possible—might recognize that the national security includes maintenance of strong basic research in all fields of quantitative science. Even without formal elimination of the relevancy requirement, such a redefinition of the DoD's mission would come close to eliminating the skewing of quantitative science basic research that we have deplored and would help reduce campus opposition to university acceptance of DoD-funded research. For the immediate future, however, some continued DoD skewing of basic research directions toward areas perceived to be of long-range military importance may be unavoidable, because congressional appropriations to DoD for its R&D may depend on actual retention of some relevancy requirements in the DoD's quantitative

86. Ibid., p. 30.

science basic research funding. Retention of those requirements for broad areas or subareas of quantitative science—so that, as now, the DoD would not fund elementary-particle physics studies or might insist that much of its atomic physics support be devoted to laser physics—though probably still poor policy, could be rational. But continued micromanagement of DoD basic research support under the relevancy criteria, as the DoD now is doing, so that, for instance, the DoD still would be deciding which specific laser physics basic research proposals really are relevant to DoD's mission, flies in the face of reason.

Funding arrangements

We next observe that allowing the DoD to directly fund university basic research invites campus dissension without any compensating benefits to the nation, whatever the merits may be of allowing the DoD to directly fund the applied research projects it prefers. Thus even if Congress will not directly fund the NSF with the basic research funds the DoD seeks, we believe the DoD would be prudent to assign those funds to the NSF for distribution and administration via the NSF's regular procedures. The fact that the DoD is diverting funds to the NSF will be publicly known, of course, as will the areas and subareas into which there has been such diversion; campus dissension will not be minimized and even may extend to NSF funding, if these diversions cause the NSF to be regarded as having been co-opted by the DoD. But as long as the DoD-diverted and originally NSF funds are commingled and indistinguishably employed by the NSF to fund research grants, many of the present campus objections to DoD basic research grants—such as have been especially voiced to Strategic Defense Initiative funding—might well be quieted.

Control of information and location of research

Irrespective of how DoD funding of university basic research is handled, DoD funding of applied research and development will continue to present profound difficulties for universities. To minimize these difficulties, universities should not knowingly accept on-campus government research support that includes any restrictions on the dissemination of research results, and they should be reluctant to accept such restrictions at the FFRDCs they manage near campus, wherein regular faculty and students are likely to be involved. Such restrictions are incompatible with the universities' mission as perceived by most students and faculty, and they put unwarrantedly severe burdens on individual researchers and the entire university community. Although universities do accept publication restrictions in grants from industry—for example, in university-industry joint-venture research, a practice we do not decry but also do not favor—principal investigators on industrial grants can negotiate the terms of such restrictions, something they are powerless to do for government grants. It also is worth noting that a principal investigator's failure to abide by restrictions in industrial grants is not a crime, punishable by stiff fines and jail sentences, as are failures to comply with secrecy or export-law restrictions.

A corollary of the immediately preceding paragraph is that universities should not permit the performance of obviously militarily related research[87]

under university auspices on or very near the campus, because such research almost certainly will have to be classified. Even if no security controls are imposed, however, the benefits, if any, to the nation in having obviously militarily related research performed on or near university campuses do not warrant the divisiveness in the university community that such campus-visible research induces, especially among the student body.

On the other hand, although universities should not administer or otherwise lend their names to government-supported on-campus research whose results cannot be freely disseminated, university faculty should be free to conduct any lawful research they choose, even if defense-related or classified, via whatever consulting arrangements the university normally permits. The willingness of many university faculty members, especially in the quantitative sciences, to accept paid consultantships from the DoD for work on defense-related research has received some severe criticism.[88] It may be that universities should forbid all consulting by their faculties as an undesirable diversion from faculty members' university responsibilities. But as long as consulting is permitted, no university that values our American freedoms should

---

87. We use the phrase "obviously militarily related research" rather than "weapons research" because much DoD-supported research—for exam-

ple, the aforementioned SEI project on reducing pilot error in targeting—has obvious military application but cannot be identified with any specific weapons system.

88. To illustrate the kind of criticism that has been expressed, with which we do not necessarily agree, we cite Marchant, "Political Constraints." He quotes physicist Michio Kaku, author of "To Win A Nuclear War," as saying that he knows scientists who double their salaries by consulting for the Pentagon and that such scientists are "like prostitutes who sell themselves to the highest bidder—only they are selling their brain instead of their body." See also Ross Flewelling and Charles Schwartz, "Resurgent Militarism in Academia," *Science for the People,* July-Aug. 1981, p. 5.

deny normal privileges to any faculty DoD consultant because some segments of the academic community have protested the content of the consulting, especially when a national majority apparently favors continuing contacts between the military and academic communities.[89]

The foregoing remarks have not criticized university management of defense-related FFRDCs located far from the campus. Good reasons can be advanced for maintaining some associations between military-related research and the university community, as we have indicated. University management of defense-related FFRDCs is a legitimate means of securing such associations. To avoid losing the respect of its university community, however, and perhaps ultimately the respect of the nation as a whole, a university administration that is managing an FFRDC should not let that FFRDC become too closely identified with the university itself, for our oft-repeated reason that the missions of the university and the FFRDC are too different. Thus, because the university administration should be seeking to convince the university community and the public at large that the university's management of a defense-related FFRDC is in the public interest, not in the university's self-interest, it is especially important that the university neither actually nor seemingly have an important financial stake in the existence of the FFRDC. The admission by Cal Tech's president that he had sought military research for JPL because the lab had lost NASA funding manifested a profound lack of understanding

89. This point has been made forcefully in comments on Zasloff, "University, by Definition," about CMU's acceptance of the SEI. Ward Deutschman and George P. Millburn, Letters to the Editor, *Chronicle of Higher Education*, 25 May 1988, pp. B2-B3.

of this recommendation and apparently did cost his administration some faculty and graduate-student support.

Dependence on research grants

Universities also should avoid becoming financially dependent on their on-campus unclassified research grants. The pressing needs of quantitative science researchers for federal funding have been described in this article. From every policy standpoint—for example, ensuring the independence of top-level administrative policies, guaranteeing that basic research problems will be chosen unconstrainedly, and enabling graduate students to follow their talents and interests freely—it is undesirable that university financial needs should be added to the already heavy funding pressures that are skewing research directions and subfield growths in the quantitative sciences.

In this connection, universities should avoid defraying with soft—that is, grant—funds any portion of a tenured researcher's salary to which the university is committed; such avoidance is not required for untenured researchers, to whom the university has no permanent salary commitment. The federal funding agencies could help in this regard by instituting a policy of refusing to charge grant funds to tenured researcher's salaries. Because, however, so many universities now have grown accustomed to paying tenured faculty salaries partially with soft money—for example, MIT, which often charges a significant fraction of a tenured professor's salary to his or her grants—such a policy, if agreed upon by the federal agencies, undoubtedly would have to be phased in.

For much the same reasons as just discussed, the sources of graduate-student support should be broadened, with less dependence on Ph.D advisers' research grants. It is unfortunate that this nation has not seen fit to fund fully the graduate education of every deserving student. In the quantitative sciences, however, because of the aforementioned shortages of competent graduate students, full public funding of such students has been achieved in effect, via departmental teaching assistantships and principal investigators' support of their graduate students. Nevertheless, the fact that most graduate students in the physical sciences cannot financially survive for the period required to obtain a Ph.D. without salaries from some grant significantly increases the funding pressures on principal investigators. At the University of Pittsburgh, for example, each supported graduate student costs the grant about $17,000 per year. At the same time, the necessity for retaining support keeps graduate students from freely following their talents and interests, as has been discussed earlier.

It would be preferable to provide more fellowships or traineeships, funded either by the universities or by the federal government, so that more students could make choices independent of the availability of federal research grant support; simultaneously, principal investigators in the physical sciences could be freed from the need to provide so substantial a portion of the total graduate-student support.

We shall not further pursue these observations, which stem from our concern that ill-conceived funding of the quantitative sciences by the DoD, even if not increased above present levels, ultimately may prove seriously harmful to university quantitative science research and to our research university system as a whole. Our conclusions about the already manifested results of DoD support—namely, the skewing effects on the quantitative sciences and the influences on university administrative and recruiting policies—are more moderate than those of Forman,[90] who concentrates on physics, and of the authors in the *Science for the People* special issue.[91] We stand by our conclusions, however, which appear to be consistent with the thrust of Holdren and Green's very thoughtful but somewhat differently focused examination of the influence of military R&D support on the nation as a whole.[92]

90. Forman, "Behind Quantum Electronics," concludes that military patronage has very strongly shaped the character of the research done by academic physicists; indeed, Forman claims that physicists have lost control of their discipline. While we find much of his analysis well taken, we do not agree that physics has been this deeply affected.

91. For example, Krinsky, "Swords into Sheepskins"; Selvin and Schwartz, "Publish and Perish"; Greg LeRoy, "War U," *Science for the People,* 20(1):11 (Jan.-Feb. 1988).

92. Holdren and Green, "Military Spending," do not concentrate on universities and certainly not on the quantitative sciences.

Electronics and Computing

By LEO YOUNG

ABSTRACT: Electronics and computing are among the fastest-moving areas of science and engineering. The universities have played and will continue to play a major role in advancing the state of the art. Nowhere is it more important to be at the cutting edge of technology from the point of view of defense, and nowhere is academic research more rewarding. These areas illustrate well the cooperation that is possible between the Department of Defense and the universities when their interests coincide.

Leo Young received bachelor's and master's degrees in mathematics and in physics from Cambridge University, England, and his doctorate in electronics from The Johns Hopkins University, Baltimore, Maryland. He has published 14 books, holds 21 patents, and has authored numerous papers, mostly in the area of microwaves and electronics. He was the 1980 president of the Institute of Electrical and Electronics Engineers. Since 1981, he has been in the Office of the Secretary of Defense, where his responsibilities have included basic research and university relations.

THE interaction between the military and the universities is illustrated well in electronics and computing. On the eve of World War II, the electron had been around for less than fifty years—J. J. Thomson's 1897 experiment is generally credited with proof of its existence—an electric calculator meant a set of noisy gears driven by an induction motor, and home television was a rare luxury. Then came the war, and the military turned to the universities—and electronics took off.

THE ROLE OF ELECTRONICS AND COMPUTING IN DEFENSE

Electronics is one of the newest disciplines. After World War II, electronics became the technology that made possible the full realization of the eighteenth-century dream of calculating machines, thus giving rise to modern computer science. The postwar volume of industrial and consumer applications gradually overtook defense applications, but the latter continue to be at the cutting edge of technology. Defense research has therefore played a particularly large role in the development of electronics and computer science and in supporting that research in academia.

The impact of World War II

Engineering has always been strongly associated with the military. Indeed, the term "civil engineer" was introduced to indicate that there exists a branch of engineering that is not military. Academic science became a major factor in military affairs in World War II, in which electronics played an especially significant role. At least three contributions involving electronics and/or computers hastened the end of the war: (1) the application of computing power to break enemy codes, which greatly helped, for example, the war at sea, as in the battle of Midway; (2) the use of microwaves to detect objects through clouds and bad weather, which enabled our bombers to find enemy targets in the dark as well as in bad weather, and the related technology of sonar, which helped detect and sink German submarines at their peak effectiveness; and (3) the harnessing of nuclear energy, using its enormous destructive power to bring the War in the Pacific to a successful conclusion.

Progress was not uniform. The urgency of war encouraged intense efforts both in combat and in research, but the rush to complete for war service aggravated problems of coordination and communications in development. For example, toward the end of World War II radars were being pushed to ever higher frequencies to resolve more detail. Ten-gigahertz radars were already in operation, and the Massachusetts Institute of Technology (MIT) Radiation Laboratory was pushing to twenty gigahertz. Unfortunately, water vapor absorbs microwaves strongly at this frequency, causing atmospheric attenuation and reducing radar range. Early measurements made during the cold dry winter did not reveal this fact, and so it was not till the warm humid spring, after a lot of effort had been expended on this radar, that the work was abandoned for a more favorable frequency. Yet, only a few steps from the radar laboratory, other scientists were mapping molecular spectra and would have known about this problem and how to circumvent it, had they been consulted.

Such lessons were not lost on the research community and their sponsors. Victory at first gave impetus to dismantling the scientific establishment that had helped win the war, but the 1945 report by Vannevar Bush recommended otherwise.[1] The first U.S. government office dedicated to the support of contract academic research in science and engineering was set up by the U.S. Navy in 1946—the Office of Naval Research (ONR). It was followed in later years by the National Science Foundation (1950), based on the ONR experience, and later, the Army Research Office (1951) and the Air Force Office of Scientific Research (1952). All of these, however, were preceded in 1946 by an electronics program. More on this unique program later.

*The influence of
world competition*

World War II established the role that electronics and computers could play in the survival of nations. At the end of the war, there existed the trained labor, the industrial resources, and the market for commercial exploitation of the new technology. The United States, being the only major industrial power not ravaged by the war, benefited the most; however, the nation was lulled into a false sense of invincibility, widely shared abroad.[2] Only a decade later, the rest of the world, with lower labor costs, newer machinery, and an open American market, was catching up fast. So much for predictions. The United States remained generally ahead in technology—to a large degree because of military spending on research—but was falling behind in the ability to produce inexpensive, quality products. Electronics was particularly vulnerable. Electronic

1. Vannevar Bush, *Science, the Endless Frontier* (1945; reprint Washington, DC: National Science Foundation, 1960).
2. See, for example, Jean-Jacques Servan-Schreiber, *The American Challenge* (New York: Atheneum Books, 1968).

products and computers were widely accepted; the state of the art was advancing rapidly; return on investment was high; working conditions, attractive; opportunities, apparently limitless. Electronics became one of the technologies targeted for focused development in Japan and to some extent in Europe. The U.S. electronics and computer industry felt the heat of competition.

The apparent inability of industry, including the electronics industry, to produce inexpensive but effective and durable—that is, quality—products has become the subject of innumerable debates. Unit costs are very sensitive to production flow and quantity; product quality depends on good initial design for functionality, manufacturability, reliability, and maintainability. This is being recognized in the Department of Defense (DoD). Recent initiatives have emphasized the increased use of commercial products in defense systems—to avoid the cost penalty of even small deviations from products in quantity production—and greater use of DoD research and development (R&D) resources to improve manufacturing processes needed for defense-related products.

The state of design and manufacturing in the United States has thus received a good deal of attention lately. Unlike a consumer who finds a foreign car more suitable, DoD cannot simply order its weapons systems from abroad. Many of the tools needed to improve our domestic manufacturing base—data bases, robots, computers, software, artificial intelligence—have roots in electronics technology. Manufacturing is increasingly becoming a subject suitable for academic pursuit. The term "concurrent engineering" has recently become prominent to denote research into the integration of product design, manufacture, and support.

The war in Vietnam caused a decline in DoD expenditures for research. In the decade 1965-75, DoD expenditures for basic research, measured in constant dollars, fell by a factor of 2. Universities turned away from defense research. Foreign students flocked to American universities, especially to study electronics and computer science. Multinational companies started manufacturing electronic products overseas. Electronics technology was soon widely disseminated over the globe, and imported electronic components found their way even into American weapons systems. Aside from the national security implications, alarm was being expressed in many quarters that whole industries were in danger and that the electronics industry might go the way of the smokestack industries, with grave implications for defense. One consequence was DoD's Very High Speed Integrated Circuits program begun in 1977, but with very little university involvement. As governmental attitudes toward antitrust regulations, particularly in regard to research, became more relaxed in the 1980s, industry research cooperatives became more popular and were encouraged by the Department of Commerce. Examples in electronics and computer technology are the Semiconductor Research Corporation, which funds university research, and the Microelectronics and Computer Corporation, which is located next to a major university campus; the latest manifestation is the Semiconductor Manufacturing Technology Center, a DoD-industry research cooperative to be located near the same university campus as the Microelectronics and Computer Corporation, the University of Texas at Austin.

Electronics R&D is perhaps particularly vulnerable to fluctuations in defense support. This seems to be borne out by the following statistics. First, it has been estimated that close to one-half of electronics engineers are employed, directly or indirectly, by DoD or its contractors. Second, and conversely, a 1980 survey of DoD laboratories showed that 44 percent of their engineers were electrical or electronic, while contemporary National Science Foundation data placed the national average at a comparatively modest 18 percent.

Universities not only play a key role in training these electronics engineers; they also perform the bulk—between 55 and 60 percent—of DoD's basic research. The history of this university-DoD relationship will be traced later in this article, when we discuss the Joint Services Electronics Program.

*Interaction between
military and civilian
technological change*

A nation's economic growth, the accumulation of wealth, and the development of international trade eventually bring about the need to defend these assets—along with the resources to do so. The commitment to military muscle stimulates technological growth while it consumes national resources. That is the leitmotiv of a recent treatise by Paul Kennedy on the rise and fall of empires.[3] As the pace of technological progress accelerates, extrapolation from the past to make predictions concerning the future becomes increasingly dangerous. The present article will stick to safer ground. It

3. Paul M. Kennedy, *The Rise and Fall of the Great Powers* (New York: Random House, 1987).

will merely recount how certain technological advances made in response to
military challenges have led to considerable beneficial fallout in the civilian sector.
We shall not go on to discuss the obvious
next question, how to maximize commercial fallout from military investments in
dual-use technologies—that is a question
that is likely to be debated for a long time
to come.

Changes produced by military technological innovation, both in content and in
management, have influenced the civilian
technological enterprise throughout American history. This fascinating subject in its
entirety is too big for this article. We refer
the interested reader to a comprehensive
exposition by M. R. Smith.[4] Some specific
examples of military-to-civilian spin-offs
of electronics technology will be given in
the following section.

DEFENSE RESEARCH
IN ELECTRONICS
AND COMPUTER SCIENCE

Most university research in electronics
and computer science sponsored by the
U.S. government is supported by DoD
and the National Science Foundation, in
roughly equal proportions. Approximately one-sixth of DoD's budget for
basic research is devoted to electronics
and computer science, a little over one-
half of it—about $100 million annually—
to universities. These figures do not include nonresearch or development categories of R&D, such as very high speed
integrated circuits and semiconductor
manufacturing technology, which involve
larger sums but much smaller percentages
to universities.

4. Merritt Roe Smith, *Military Enterprise and
Technological Change* (Cambridge: MIT Press,
1985).

The Joint Services
Electronics Program

Perhaps the longest continuous research program in DoD, and one of the
most productive, is the Joint Services
Electronics Program (JSEP).[5] It celebrated its fortieth anniversary with a
symposium held at the National Academy
of Sciences in September 1986.

JSEP grew out of the National Defense
Research Committee, set up by the Congress in 1940 to establish research centers
oriented to assist the war effort by developing new military concepts and equipment. Vannevar Bush was director of the
committee. Principal academic electronics
research centers were at MIT, Harvard,
Columbia, and the Brooklyn Polytechnic
Institute.

The MIT Radiation Laboratory, under
Lee DuBridge, specialized in microwave
radar, working closely with Rudolf
Peierls's group—which included J. T.
Randall and H.A.H. Boot, among others—
at Birmingham University in England.
The Peierls group demonstrated the first
stable magnetrons in high-power radar
transmitters. Meanwhile, Rudi Kompfner
at the British Admiralty pioneered low-
noise traveling-wave tubes for sensitive
radar receivers, which he later adapted
for satellite communications when he
joined the Bell Telephone Laboratories
after the war. The Harvard Radiation

5. Much of the material on JSEP is drawn from
two excellent reports: Alton L. Gilbert and Bruce
D. McCombe, *Joint Services Electronics Program:
An Historical Perspective,* prepared for the U.S.
Army Research Office, Electronics Division, Apr.
1985; David Robb and Arnold Shostok, eds. *Pro-
ceedings of the Fortieth Anniversary Symposium of
the Joint Services Electronics Program,* for the U.S.
Air Force Office of Scientific Research, the U.S.
Army Research Office, and the U.S. Office of Naval
Research, Jan. 1987.

Laboratory, led by Fred Terman, concentrated on electronic warfare and countermeasures: jamming and deception of enemy radars and communications. When Terman's group returned to Stanford University after the war, they started a JSEP there. The Varian brothers had experimented at Stanford before the war on a versatile microwave tube called the klystron; it was further developed and manufactured by the Sperry Corporation during the war. The Columbia Radiation Laboratory, under the direction of Professor Isidor Rabi, developed the highest-frequency—that is, shortest-wavelength—microwave components, prepared as they were for it by Rabi's prewar research on microwave spectroscopy. The Brooklyn Polytechnic Institute's Microwave Research Group, under Ernst Weber, concentrated on microwave measurements and passive components. It was a time of close cooperation between universities, industry, and government.

When the National Defense Research Committee was disestablished after the war, the need for continued coordination of research in electronics was recognized by the military services. A DoD committee was appointed in early 1946 to phase out the radiation laboratories; it consisted of Lieutenant Colonel Harold Zahl (Army), Commander E. R. (Manny) Piore (Navy), and Major John Marchetti (Army Air Corps), all of whom had been engaged in research with Norbert Wiener at MIT during the war. They recognized the need for DoD to continue peacetime research, and they had to overcome a number of bureaucratic hurdles in pursuing this goal. One even required the personal intervention of President Harry S Truman at the request of then Columbia President Dwight D. Eisenhower, who was enlisted by Professor Rabi after

other efforts had stalled. Thus the tri-service JSEP effort grew out of the wartime experience of the radiation laboratories and benefited from the personal connections forged by the war. Many distinguished scientists and engineers accepted the call to be JSEP managers, including Nobel laureates Isidor Rabi, Charles H. Townes, Polykarp Kusch, and Nicolaas Bloembergen. Other electronics researchers who received recognition through the Nobel prize and who were supported at some time or another under JSEP or another DoD program include John Bardeen, Leon Cooper, Leo Esaki, Willis E. Lamb, Arthur Schawlow, and John Schrieffer. Many other major awards resulted from JSEP-sponsored research. The list of participating universities has also grown. Current membership includes MIT, Columbia, Polytechnic University (Brooklyn), Harvard, Stanford, Ohio State, Cornell, the Georgia Institute of Technology, and the Universities of Illinois, California at Berkeley, Southern California, and Texas at Austin.

To keep JSEP in perspective, it should be noted that the program was never more than a small fraction of the DoD-supported research in electronics. For example, the first electronic—as opposed to electromechanical—computer, the Electronic Numerical Integrator and Calculator (ENIAC), was built by Professors Eckert and Mauchly and their coworkers at the University of Pennsylvania for the U.S. Army Ballistic Research Laboratories. The Ballistic Research Laboratories needed it to calculate shell trajectories and firing tables. Started in 1943, completed in 1946, it had a useful life until 1955. ENIAC did not have a programmable memory like modern computers. Instead, engineers had to change wire connections to reprogram. But the concept of the general-purpose stored-program computer was first published by John von Neumann in a draft report under the ENIAC project. Similarly, other DoD establishments have always sponsored research in pursuance of their objectives, often at institutions the better prepared for it as a result of JSEP.

Each JSEP contract focuses on a broad area of technology corresponding to local expertise and interests. It permits more than the usual autonomy to the local program manager within his or her technology area. This flexibility makes possible a faster follow-up on scientific breakthroughs and permits some risk taking within each program.

Organizationally, JSEP is run cooperatively by the three service research organizations, that is, the Air Force Office of Scientific Research, the Army Research Office, and ONR; regular on-site reviews are held by the tri-service JSEP committee in the presence of other JSEP university representatives. This cooperation fosters informal and rapid information transfer between the services as well as between the participating universities. Other DoD agencies, such as the Defense Advanced Research Projects Agency, participate actively in these exchanges.

JSEP is relatively small, running on about $10 million per year overall. It represents only about one-tenth of all DoD research in electronics and computer science at universities. Its influence has nevertheless been great, not only in the historical context but also because it has been well managed as an integrated and productive program based on an interactive team that communicates and shares results rapidly. JSEP has provided continuity and has served as a forum and as a focal point, thus facilitating long-term

planning. It also has made seminal contributions—published without restriction.

Military R&D has made significant contributions to civilian spin-off products and even to whole industries. Thus the Stanford Electronics Laboratory, funded mostly by military R&D, influenced the growth of the electronics industry around the university that turned into Silicon Valley. Professor John Linvill, performing JSEP research on the Stanford campus, was interested in refining the Optacon, a device to enable blind persons to read printed material. The thrust of his effort was to develop miniaturized microelectronic circuits, and it coincided with similar goals for military airborne and space-type hardware.

At the JSEP-supported Electronics Research Laboratory of the University of California at Berkeley, Professor Donald Peterson and his students undertook an active program in computer-aided design of integrated circuits and their fabrication. One modeling program developed by them and widely used in the electronics industry was called "Simulator Program with Integrated Circuits Emphasis."

JSEP contracts are large enough to permit a high degree of autonomy within the overall thrust of each university program, thus facilitating interdisciplinary research. JSEP served as a model for the University Research Initiative Centers started in 1986, which had as one of their primary goals the encouragement of interdisciplinary research.

A few months after JSEP was started, an act of Congress established ONR to continue coordinated support of research activities in all defense-oriented disciplines. Soon afterward, ONR was followed by the establishment of similar research offices in the Army and the Air Force.

Electronics research in DoD today

JSEP represents a microcosm of electronics research. This characteristic and its long, distinguished history made it instructive to present it here as a special case study. DoD supports technologies required in search, surveillance, tracking, guidance, communications, navigation, command and control, electronic warfare, and other areas; these technologies require electronics for high-speed information processing, high reliability and availability, operation in hostile environments, and resistance to jamming and interception. DoD electronics research explores the electromagnetic spectrum from radio to optical frequencies, including antennas and propagation, transmitters and receivers; it addresses complex and unpredictable situation management, personnel training, measurement techniques, diagnostics, and maintenance of equipment. The program is structured in a logical progression from electronic and electro-optic materials to electronic devices, to components, circuits, and systems.

Defense research has a long list of accomplishments in electronics and computers, both hardware and software. It has played a major role in the development of navigation: Loran for radio navigation, developed at Harvard; ring laser gyros for inertial navigation, developed at Columbia, MIT, and Stanford and based on the maser and laser developed at Columbia; microwaves and millimeter waves (MIT, Polytechnic, Harvard, Stanford, and Georgia Tech); antennas and phased arrays (Harvard, Polytechnic, and MIT); solid-state circuits (Stanford and Berkeley) and electronic materials (Illinois, Texas, and Stanford);

nonlinear optics (Harvard); microwave acoustics (MIT, Stanford, and Polytechnic); information and coding theory (MIT and Berkeley); the Fast Fourier Transform (Princeton), and digital signal processing (Princeton and MIT); widely used software models for the computer-aided design of integrated circuits (Berkeley and Stanford); computers since ENIAC (Pennsylvania); superconductivity (Illinois and Stanford); plasma displays (Illinois); and so on. This list is far from complete; nor are the universities mentioned the only ones to have made major contributions.

The unique and growing demands made on computers provide strong motivation for an active program of basic research in electronics and the computer sciences. DoD has provided leadership in the development of time-sharing, higher-level languages, packet-switched communications and distributed processing, artificial intelligence, computer-aided design and computer-aided manufacturing, robotics, multiprocessor architectures (supercomputers), interactive computer graphics, man-machine interactions, embedded computing, and microelectronics architecture and software design tools.

The DoD program in electronics and computer research is a mix of contracts and grants, with various technological and programmatic objectives. Some of the program is devoted to the purpose of increasing fundamental understanding. Some of it has the goal of exploiting new technologies for a plethora of applications such as sensing, communications, signal analysis, cognition, decision making, and neural bases, as well as in modeling, simulation, design, manufacturing, automation, robotics, diagnostics, and logistics. Some of it is to encourage interdisciplinary research, which is especially important in these rapidly evolving technologies.

Programmatic considerations in defense research

There are many shifting needs and issues in research that apply in differing degrees to all disciplines and technologies. The following remarks therefore apply not only to electronics.

One of the issues is fundamental versus directed research. It suffers from a lack of precise operational definitions and is therefore difficult to quantify. The issue was aptly addressed in a comment, made by Charles Townes, that only the former could have produced the laser, while the latter is more suitable to improving the light bulb. Nevertheless, a case can be made for pointing research in particular directions. Thus it is appropriate for DoD to support extensive research in computer science and engineering but to leave most research in the health sciences to the National Institutes of Health. The exceptions would include diseases rare in the United States but common in certain parts of the world, contagious diseases, and wounds inflicted in battle.

A related issue is that of basic versus applied research. These two concepts also lack satisfactory operational definitions. They often differ more in intent than in content. A concept that came into vogue a few years ago was that of focused research, which may be defined as a group of research tasks focused on an application area of special current interest, say, antisubmarine warfare. This encourages interdisciplinary research efforts. In the case of antisubmarine warfare, such efforts might be a combination of research in acoustic propagation, noise reduction, and improved acoustic sensors. This approach provides management with a tool for channeling research thinking—intent—into areas of special interest without interfering with the research process, the content.

How much research should be placed in the hands of individual researchers compared to relatively large research centers? The latter implies greater local management responsibility and accountability and hence a certain degree of institutional autonomy. The former often hinges on one individual's talent or expertise, while the latter is better suited for multidisciplinary research, where team work counts as much as originality.

Should DoD research dollars pay for investments such as education and training, instrumentation and facilities, the residue of which accrues to the performing organization or individual, in addition to consumables like time and matériel, which are more visibly for the public benefit? Where should the line be drawn? This question will continue to vex. Current needs and shortages tend to dictate the right mix; there is no universal algorithm to replace human judgment.

Cooperative research is growing, partly in reaction to earlier, too stringent interpretation of antitrust regulations applied to research; partly as a result of successful cooperative endeavors in Japan, which, however, may be inappropriate models for the United States; and also because of a wider recognition that all parties to a cooperative endeavor may gain from pooling resources. There is a developing consensus that academia, industry, and government must collaborate more closely in a changing, highly competitive world.

Important considerations for DoD-funded research are technology transfer and technology transition. Technology transfer may be defined as the dissemination of scientific or engineering information. It is the process of moving technology laterally, from one performer to another. Technology transfer is best accomplished by direct collaboration among scientists or engineers; it is accomplished less effectively through a third party or instrument, such as publication in a journal. There are many factors that inhibit technology transfer, such as proprietary and patent considerations, national security issues, and scientific priority, or not wanting to be scooped, and so forth. Universities are in an ideal position to play a major role in technology transfer. Thus a cooperative research center on a university campus with industrial participation can greatly aid technology transfer.

Technology transition may be defined as the first embodiment of a new technology in a useful product. It is the process of moving technology vertically, from theory to practice. Marketing a new technology can be more difficult than marketing other products because it is often hard for the user or consumer to appreciate the potential of a new and perhaps esoteric technology. Effective technology transition requires the active cooperation of user and producer. An advanced-technology-demonstration project could serve as a vehicle for such cooperation. Universities could play a larger role in proposing as well as in planning such projects.

FUTURE DIRECTIONS

Electronics and computer science deal with information, its sensing, collection, processing, storage, organization, retrieval, communication, management, control, and usage. If the first industrial revolution was based on new sources of energy and the machinery that made them possible, the second industrial revolution will be based on the vastly more efficient organization of resources by machines that process information, not energy.

A growing proportion of workers are engaged in the information industry. Information is vital in business and trade, in manufacturing and service, in communications and transport, and wherever people and systems interact, including in defense. The modern world could not function without speedy and efficient information handling by electronic means.

In addition, there remains much basic research to be done in analog, or nondigital, electronics, from microelectronics to high-power electronics, from molecular to biological electronics, from microwave to infrared wavelengths and beyond. Will there be superconductors operating at moderate power levels at microwave and millimeter-wave frequencies? To what extent will silicon be replaced by compound semiconductor materials? Will optical waveguides replace metallic conductors on integrated circuits? Will analog logic circuits become competitive with digital ones, at least for some superfast logic applications? There are already many research applications of electro-optic, electro-acoustic, and magneto-elastic devices, and they are growing. Applications of electronics to robotics, manufacturing, and other fields offer tremendous opportunities.

Knowledge-based systems

Just as machines have amplified the effectiveness of human muscle, so computers have supplemented, but not supplanted, the human brain. Machines can perform tasks undreamed of a generation ago. Early attempts often mimic familiar objects, but they gradually evolve into something more distinctly adapted to the task at hand. Automobiles evolved from horseless carriages, and early robots imitated people, but there the similarity ends. Unlike a person, a robot can perform a given task time and time again with the same precision, without tiring, and can turn out products faster and more consistently. Sophisticated programs may allow for a variety of responses depending on external stimuli.

Computers do more than number crunching. They can be made to handle symbolic logic; they can be endowed with reasoning power. In a rudimentary way, such is the case with so-called smart weapons that find their own targets after launch, or with advanced numerically controlled machine tools that find the next part to pick up, process it, and pass it on. Some programs may permit the machine to learn from experience. Eventually, we may automate some decision making. We have entered the fascinating realm of artificial intelligence. There are many applications: pattern recognition, such as finding the target in the scene, or the flaw in the manufactured part, or a tool on the factory floor, or even a legal precedent; diagnosis of a disease, or a faulty engine, or a failing bridge, from the symptoms; and prognosis such as forecasting the weather, or the producibility and maintainability of a proposed new product or design. Prognosis is more subtle than diagnosis since the latter—working backward through deductive logic—ultimately should yield only one correct answer, while the former—working forward and making predictions—is more probabilistic in nature, with alternative answers depending on the assumptions made. For instance, how will the engine be used? in what climate? in what facility will it be serviced?

Vast amounts of information may be stored in data bases known as knowledge-based systems or expert systems. These are data bases that have captured and organized information provided by experts and that can subsequently be brought to bear systematically on a problem. The information may be processed by an inference engine, which makes decisions based on predetermined rules. The fundamental question is, To what extent can human judgment and perhaps human values, be programmed in? The answer will change as technology moves on.

The role of the universities

Every military technology has its counterpart in the world of manufacturing or medicine or business or finance. Military technologies often stimulate similar research problems. What should DoD's role be among sponsors of academic research?

From the defense point of view, universities are not only a great resource whose ideas have been literally proven in battle, but they also represent an invaluable window on the world in peacetime. Military plans by their very nature cannot be published freely, yet the ideas and concepts of science and technology cannot be kept secret for long. By supporting academic research, DoD ensures its scientific currency, and the universities gain more than funding: they are challenged to solve some of the toughest science and engineering problems of the day, often lending themselves to unforeseen applications in industry and elsewhere. Furthermore, as educational institutions, universities educate and train the next generation of engineers and scientists, who must bear the responsibility for continued progress.

High-technology applications in defense have in the past proven to be powerful stimuli for university researchers. The challenge now is to maintain open communications and mutual respect between the universities and DoD not only under the threat or actuality of war but also in times of peace, enhancing technological progress by developing dual-use technologies and at the same time providing the best insurance for peace. And that is no mean challenge.

DoD, Social Science,
and International Studies

By RICHARD D. LAMBERT

ABSTRACT: A general cycle of relations between the Department of Defense (DoD) and the university is described with particular reference to the social sciences and international studies: a general decline in amity since World War II, decreased support for DoD objectives, a concern for the effect of DoD priorities on the general research profile, the growth of in-house and nonacademic vendors in research and training, and the enclaving of the military-connected research community within the university. The pattern of DoD support for strategic studies, linguistics, and language and area studies is examined.

Richard D. Lambert is emeritus professor of sociology from the University of Pennsylvania and director of the National Foreign Language Center at Johns Hopkins University. He is past president of the Association for Asian Studies and of the American Academy of Political and Social Science, whose Annals *he now edits. His most recent publications include* Beyond Growth: The Next Stage in Language and Area Studies; Points of Leverage; *and* The Transformation of an Indian Labor Market.

RELATIONS between higher education and the Department of Defense (DoD) are convoluted and stressful. To paraphrase Dwight Eisenhower, the military-educational complex needs constant attention. Increasingly, the question is being raised on both sides of that relationship as to whether it is currently operating in a fashion that fully serves our national needs. For one thing, the relationship is becoming inordinately stressful. Small things often raise hackles on both sides of the divide. Nowhere is this situation more apparent than in relations between social science and the military, particularly in those aspects of social science that illuminate international affairs. One would have thought that since our military's mission largely has to do with operations in other countries, the various branches of international studies on American campuses would nurture and inform that mission, whether directly or indirectly, and that, conversely, DoD would have a vital stake in assuring the health and vitality of international studies. Neither appears to be so. It is to this interface between the social sciences and the military—in particular, academic international studies and the military—that we direct our attention.

HISTORICAL BACKGROUND

First, consider a bit of the general recent history of university-military relations. For the social sciences, the military connection in both world wars spurred important growth. In World War I, major advances were made in the measurement of intelligence. In World War II, key social scientists moved from positions of external advisers to internal policymakers conducting research intended to influence policy directly. For instance, sociologists, anthropologists, and political scientists

worked in the Research Branch of the Information and Education Division of the War Department and in the Foreign Morale Analysis Division.[1] Out of this experience came major developments in attitude measurement, scaling, the fit of individuals to organizations, the role of values in behavior, small-group research, and many aspects of social engineering. Views on scientific management were carried from the factory to the military. Experience gained in the management of Japanese internment camps, in the training of Office of Strategic Services personnel, in the creation of effective command structures among combat troops, and in the conduct of military occupation generated theories on the governance of people more generally. At the end of the war, it was social scientists who developed the point system that dictated the order in which people were released from the services.

Both before and throughout World War II, the scholars who participated in the military effort had strong connections with major research universities and with national research organizations such as the Social Science Research Council. In the postwar period, many of the social scientists who had worked in the military moved back into academia, where they established new styles of research and encouraged a major new emphasis on theory as the basis of scientific advancement. Three of the founding members of the influential new Harvard Social Relations Department had served in the military social science establishment: Clyde

1. For an excellent analysis of the role of influential social scientists in the military during World War II, see Peter Buck, "Adjusting to Military Life: Social Scientists Go to War, 1941-1950," in *Military Enterprise and Technological Change,* ed. Merritt Roe Smith (Cambridge: MIT Press, 1985), pp. 203-52.

Kluckhohn, Henry Murray, and Samuel Stouffer. By 1967, seven scholars actively involved in wartime military research had served as president of the American Sociological Association.[2]

Some of the work of the social scientists spilled over into what became known as area studies. Clyde Kluckhohn was the first director of Harvard's Russian Research Center. During the war, many were involved in the study of other societies, for instance, in attempts to assess the morale of the Japanese population toward the end of the war, and in a major survey in Germany immediately after the war in an effort to measure the effect of strategic bombing on the morale of the civilian population.

These two wartime exercises of social scientists also indicated the limits of the impact of social science research on operational policy in the military. The findings of social scientists that Japanese civilian morale was declining and thereby might undermine the national will to win the war without a major destruction of civilian targets by the Americans was ignored. The evidence of the Strategic Bombing Survey[3] that saturation bombing had very little effect on Germany's willingness to fight was ignored at a later date, when the military decided to utilize the same approach with respect to North Vietnam. Similarly, several decades after the survey, when a group of distinguished political

2. Recently there has been a revival of interest in this period. For instance, an unpublished manuscript by Talcott Parsons on the relation of social science to national policy has just appeared. See Samuel Klausner, *The Nationalization of the Social Sciences* (Philadelphia: University of Pennsylvania Press, 1987).

3. "The Effects of Strategic Bombing on German Morale," in *The United States Strategic Bombing Survey* (Washington, DC: Government Printing Office, 1950). See also David MacIsaac, *The Story of the U.S. Strategic Bombing Survey* (New York: Garland Press, 1976).

scientists addressed a letter to President Nixon urging the cessation of the bombing of Cambodia, the impact of that letter on military policy was negligible at best.

Indicative of the fuzzy civil-military boundaries of the time is a little-remembered episode that took place between the end of the war in the European theater and its successful completion in Asia. As a token of the high value given to advanced education, the U.S. Army set up from scratch a complete, fully staffed American university in Biarritz, France, to assure that members of America's future intellectual elite could resume their college studies while they served the remainder of their enlistment in Europe. This blurring of institutional boundaries raised no symbolic issues at the time. Today a hostile cannonade would be happily booming along.

As a further indication of the easy mix of military and civilian purposes during the war, the creation of the Army Specialized Training Programs and the establishment of Navy V-12 programs on American campuses were as much attempts to support those hard-pressed institutions whose students were all off to war as it was to train engineers and area studies personnel for whom demand was at best uncertain. In the years immediately after World War II, university campuses performed all kinds of military research without discomfort.

Moreover, past military service was a major supporter of students in universities. A substantial portion of the wave of student enrollments that made the universities' rapid expansion possible in the 1950s and 1960s was composed of veterans who were supported in their college training by the GI bill. Without that influx, it is difficult to believe that the immense expansion of higher education in America, particularly in our graduate schools, would have occurred. In addition, over the years, Reserve Officers' Training Corps programs were established on many campuses. By 1986, there were 534 universities and colleges with these programs, with 92,786 students enrolled.[4] It was in this context of relatively amicable relations that the military services began a long-term process of supporting on-campus research in areas of shared interest. The Office of Naval Research was established by Congress in 1946, the Army Research Office in 1951, the Air Force Office of Scientific Research in 1952. All of them were dedicated to the support of external applied and basic research, much of it to take place on university campuses. Even the creation of the National Science Foundation, as indicated in its initial statement of purposes and governance structure, had a legislatively mandated role for the military. Its first director and many of its senior staff came from the Office of Naval Research.

In the following decades, several things happened. First, there was a period of demobilization immediately after the war—we went from 12.0 million troops to 3.0 million in less than a year and by 1948 we were down to 1.5 million. This figure, however, was still 640 times that of prewar levels. Many pressures pushed the scale of our military establishment upward: for instance, participation in the Korean War and the Vietnam war, the fact that the United States was the only major industrial state to escape relatively unscathed from World War II and therefore assumed a position of leadership in confronting the perceived Soviet challenge both in manpower and in technology, and the negotiation of treaties with other countries that committed us to an extended military posture, including the permanent stationing of troops abroad. Manpower levels peaked during the Korean and Vietnam wars but in other times plateaued at about 2.0 million, far above the notion of the small standing army that had been the United States' tradition throughout most of its history.[5]

It was not the growth in manpower, however, that was most influential in determining the relationship of the military with the universities; it was the need to advance rapidly in research and development in weapons technology. To achieve and maintain the technological superiority on which American military strategy was based, a major research-and-development effort had to be made, one that required external as well as internal research. Military funding of research and development in technology and engineering increased from $5.4 billion in 1960 to $33.6 billion in 1986.[6] To be sure, most of DoD research moneys do not go to universities but to industrial contractors in the private sector. In 1986, out of the total investment of $33.6 billion, only 3.2 percent, or $1.0 billion, went to universities. The universities' share of what is termed basic research, as contrasted with applied research, is considerably larger: 54.5 percent of DoD funds for basic research in 1986 went to universities.[7] Of course, other sections of the government were also expanding their investment in campus-based research. The federal government accounts

4. Correspondence with Colonel David M. Beckman, director of the Education Directorate, DoD.

5. For an account of this transformation, see Christopher J. Deering, "Congress, the President, and Military Policy," *The Annals* of the American Academy of Political and Social Science, Sept. 1988, vol. 499.

6. U.S. Department of Defense, *Report on the University Role in Defense Research and Development,* prepared for U.S. Congress, Committees on Appropriations, Apr. 1987, tabs. 1 and 3.

7. Ibid.

for two-thirds of research and development funding at universities. In 1986, DoD funding accounted for only 16.4 percent of all funds from federal agencies for university-based research, but a billion dollars a year flowing from the military to the universities for research cannot help but have a major impact.

The implications of this increase in funding and the increasing dependency of university-based natural science and engineering research on DoD has concerned many thoughtful spokespersons.[8] They worry about the skewing of our national research priorities to military ends, and the movement of the center of decision making for our national research effort to offices in DoD. They worry about the effect of making the research enterprise dependent on a contract versus a grant-funding mechanism, of merit versus peer review[9]—in particular, the intrusion of geographic and other nonscientific considerations into the selection process—of the assignment of priority to narrow, short-term, product-oriented research rather than longer-term, more basic research, and, above all, of the utilization of the fruits of scientists' research for policy objectives of which they disapprove.

This increasing unease in the civil-military research interface growing out of such organizational issues was accompanied by a growing alienation of impor-

tant sections of the university-based intellectual community from our nation's foreign policy, at the same time that that policy became increasingly intertwined with military policy.

Particular events can be picked out to mark the progress of the alienation—the McCarthy era, although the Army was one of the principal victims; the antinuclear movement; declining civilian support for the overlong Korean and Vietnam wars, and reaction against what was perceived as the misuse of social science in those wars; our military policy in Central America; and, more generally, the growing use of clandestine operations to achieve foreign policy ends. In general, except perhaps for the Berlin Airlift and the Cuban missile crisis, the easy sharing of purpose between the military and academia that was so evident in World War II and the years immediately afterward gradually evaporated.

At the same time, three other processes were under way that had major effects on military-university relations. First, the military developed its own training facilities and expanded its research activities in-house, even though the universities' share in the total technology base and basic research programs expanded. A little-appreciated but dramatic illustration of this is the fact that there are now 38 accredited postsecondary schools maintained by DoD, including 21 that provide both graduate and undergraduate training and 17 that provide specialty training within particular services. While campus-based Reserve Officers' Training Corps programs continued, the more military training—particularly the more technical aspects of military training—were moved in-house. DoD found its needs not to be adequately served by the civilian formal educational system, so it set up its own. Similarly, in support of

8. See, for instance, U.S. Congress, House, Committee on Science and Technology, *Hearings on Science and the Mission Agencies,* "Testimony of Rodney Nichols before the Task Force on Science Policy," 23 Oct. 1985.

9. For the Department's viewpoint, see U.S. Department of Defense, *The Department of Defense Report on the Merit Review Process for Competitive Selection of University Research Projects and an Analysis of the Potential of Expanding the Geographical Distribution of Research,* prepared for U.S. Congress, Committees on Appropriations, Apr. 1987.

research, in 1986 DoD spent 32 percent of its basic research and 42 percent of its applied research moneys intramurally.[10]

Second, DoD increasingly dealt with industrial firms, with competitor quasi-academic organizations such as Rand and SRI International, or with one of the host of so-called beltway bandits who could apply for, and deliver, research in the format that DoD favored. DoD-supported research by industrial firms constituted about 30 percent of the total, and another 3 percent was performed by nonprofit institutions other than universities and colleges. Together, the in-house and the non-university-based external research establishment grew immensely. DoD Research, Development, Test, and Evaluation funds expended by nonacademic institutions, including their own laboratories, increased from $3.9 billion in 1955 to $32.6 billion in 1986.[11]

Third, within the university world itself, military-university relations has become a specialized function of an enclaved, limited subset of organizations and individuals. On the one hand, the large contract research shops like the Applied Physics Laboratory at Johns Hopkins University manage DoD research on a large scale. The various services maintain 10 Federally Funded Research and Development Centers located both on and off the campus that serve as principal research contractors that receive both general and preferential treatment in project support. Very recently, DoD has shown its preference for dealing with organized campus-based research organizations in its plan to use the moneys in the congressionally mandated University

Research Initiative to establish additional DoD-related research centers on campuses around the country. On the other hand, a relatively small number of individual specialists have become principal participants in DoD-sponsored research, leaving the rest of the professoriat largely untouched. DoD's own estimate is that it supports 14 percent of the nation's scientists and engineers in universities and industries. As we will see, this delimiting of DoD contacts among the professoriat has especially important implications for international studies.

THE SHARE
OF SOCIAL SCIENCE

With one exception, the general historical sketch outlined also characterizes the relationship between social scientists and DoD: a general decline in amity since World War II, a decline in support for DoD's objectives, a concern for the effect of DoD priorities on the general research profile, the growth of the role of in-house and nonacademic vendors in research and training, and the enclaving of the military-connected research community within the university. The exception to that general history is the limited funding that DoD provides for the social sciences in comparison with the natural sciences and engineering.

Determining DoD's expenditures on social science research is no easy matter. One source that reports on general federal support of research by agency states that, in 1986, only $60,000 was obligated for research in the social sciences performed at universities and colleges.[12] A more detailed report focusing only on universities, colleges, and nonprofit organizations indicates that $2.292 million was

10. U.S. National Science Foundation, *Federal Funds of Research and Development: Fiscal Years 1986, 1987, 1988,* NSF 87-314, vol. 36, 1987.

11. Department of Defense, *University Role in Defense Research,* p. 7.

12. National Science Foundation, *Federal Funds for Research and Development,* tab. C-92.

spent for that purpose.[13] This figure does not include classified research supported by the intelligence agencies—for instance, the Defense Intelligence Agency (DIA) spends about $500,000 annually in support of external social science research and consultation—and some forms of project research that fall under the operations and maintenance portion of the services' budgets are not reported. Nonetheless, it gives some indication of scale. The figure $2.292 million represents 2.0 percent of the total DoD expenditures for research at academic institutions. It includes economics, political science, sociology, and assorted social science topics. Anthropology and history are reported as receiving no support at all. If one adds social psychology to the list of social sciences, another $11.754 million was given to academic institutions for research, bringing the total up to 12.2 percent of all expenditures.

This pattern of funding is an indicator of DoD's general estimation of the social sciences. By and large, it shares the perspective of most engineers and natural scientists: it has little interest in them. Moreover, DoD feels bound by the Mansfield Amendment to support only research of demonstrable relevance to the military mission. It is difficult for social science to meet this criterion. When the pattern of expenditures under the new DoD University Research Initiative—which was specifically designed to support the university-based research infrastructure—was being discussed, at least one formal representation was made to DoD to extend the coverage of that program beyond the natural sciences and engineer-

ing. The suggestion drew no support whatsoever.

While DoD has no particular interest in the social sciences, most social scientists are generally indifferent or hostile to DoD. Some of this distancing resulted from differences in notions of what rules should govern the relationships between scholars and the uses to which their research is to be put.[14] The most celebrated of these misunderstandings, which resulted in an upsurge of moral outrage in the social science community, was Project Camelot.[15] This was a DoD-funded large-scale research project on "the causes of revolutions and insurgency in underdeveloped areas of the world." The precise source of the outrage was that several scholars in Latin America had been paid from DoD funds without their knowledge of the source. This issue of the acceptability of DoD funding to scholars abroad has remained a central source of friction within international studies, as we shall see. It also precipitated a rancorous debate, part of which had its roots in the uses of social science research in the Vietnam war, about the acceptability of DoD as a bedfellow under the best of circumstances.

As the years went by, most social scientists saw themselves as public critics, not formulators, of American foreign policy in general, and the role of the military was viewed as only one part. The relationship between social science and

13. U.S. National Science Foundation, *Federal Support to Universities, Colleges and Selected Non-Profit Institutions: Fiscal Year 1986*, NSF 87-318, tab. B-14.

14. For one report on the difficulties an academic felt in dealing with DoD's research support in the 1960s in Southeast Asia and elsewhere, see Seymour Dietchman, *The Best Laid Plans* (Cambridge: MIT Press, 1976).

15. See K. H. Silvert, "American Academic Ethics and Social Research Abroad," *American Field Service Reports*, 1968; Irving Louis Horowitz, "The Life and Death of Project Camelot," *Transactions*, 3:3-7, 44-47 (1965).

DoD settled into a posture that can best be described as disengaged and wary.

DoD AND
INTERNATIONAL STUDIES

Of special interest is that section of the social science community that deals with foreign affairs, the domain in which DoD policy is supposed to operate. One might expect that the academic base for expertise on other parts of the world and of the global political and economic system would be of direct interest to DoD planning. How does campus-based international studies relate to our military system? The answer is that the same pattern of amity to estrangement characterizes this subset of relations as well, with several special twists.

To begin with, international studies includes within it a series of specialties that differ in their relationship to DoD. We take as examples three such specialties: (1) international relations and strategic studies; (2) foreign language education; and (3) language and area studies. By and large, these specialties have remarkably little to do with each other, and each has a different history of dealing with DoD.

International relations
and strategic studies

The study of relations between countries has a long and distinguished academic history. In the period immediately after World War II, such studies tended to concentrate on issues of foreign policy—ours and those of others; on the newly developing United Nations; and on the general properties of the world political system. The results of such research were more interesting to the State Department than to DoD, which was in any

event busy dismantling the huge military machine created for World War II. In the late 1950s and 1960s, studies of international relations expanded to include research on decolonization, development, and the emerging Third World. By and large, these, too, were civilian, not military, concerns. There were, however, American academics concerned with strategic policy.[16] This interest gradually expanded, spurred in part by the cold war and concern for nuclear strategy. Academic specialties developed in strategic studies, arms control, and, more recently, peace and security studies, all of which were closely linked to military policy.[17]

As the field grew, the expansion in our defense establishment, noted earlier, was taking place. It was accompanied by a sizable increase in the funds available for support of research. The same pattern of interaction that characterized general DoD-university relations was characteristic of interaction between DoD and international studies as well. DoD employed as external researchers first individuals, then campus-based research centers, and then freestanding contract research shops, sometimes directly connected with the military.[18] The Rand Corporation's symbiotic relationship with the Air Force is only one important example.[19]

16. See Fred M. Kaplan, *The Wizards of Armageddon* (New York: Simon & Schuster, 1983).

17. For a review of this history, see Robert O'Neill, "A Historical Perspective on International Security Analysis," *Items,* June 1988, p. 42.

18. For a list of some of these grantees, see National Science Foundation, *Federal Support to Universities,* tab. B-33.

19. It is interesting to note that when the Rand Corporation broadened its research focus, its style of research with its emphasis on quantitative analysis and rigorous research design served as a prototype of academic public policy research in general.

Early DoD-commissioned research tended to be narrowly focused on particular strategic problems or on specific weapons systems. As the field of strategic studies matured, a systematic, highly quantitative style of analysis of organized conflict game theory, conflict simulation, and so on developed within the academic community and had an immense impact on military planning. Gradually, the focus of international relations research broadened to include an examination of grand strategic issues. At the same time, in part to privatize strategic planning research, several private foundations—notably, Ford and Rockefeller—nourished the growth of an independent interdisciplinary academic field of strategic studies. More recently, in part through the intervention of the MacArthur Foundation, the emphasis in strategic studies shifted from how to make war to security studies with an emphasis on how to increase the likelihood of peace.

In the course of this transition, DoD's relation to the field shifted. It changed from being a principal patron to being a distant, occasional consumer of the product of these studies. While many of the results of academic strategic planning research were directed to military policy, it is unclear how much attention planners in the Pentagon actually paid to the recommendations emerging from that research. Casual evidence indicates not much. Meanwhile, DoD proceeded to move much of the strategic planning research in-house, using academics as advisers, or, when it dealt with outsiders, contracting out to its customary client research organizations.

Foreign language education

One area that did, and still does, receive a fair amount of DoD support is linguistics, particularly applied linguistics as it relates to foreign language instruction. During World War II, the army developed oral-aural texts and teaching materials that became part of a major revolution in foreign language instruction. Teaching materials were prepared in languages that most people had never even heard of in earlier years. After the war, several of the divisions of DoD were among the primary sponsors of an interagency effort coordinated by the National Science Foundation to develop machine translation of foreign languages. This project, begun in 1953-54, funded eight research centers on American campuses. After about a decade of inconclusive research, these projects were discontinued, but in their place a National Science Foundation program first in computational linguistics, then in general linguistics, was created, and this program serves as a major source of support of academic research on linguistics today.

More recently, the National Security Agency (NSA) has quietly served as the principal source of funds for research on the improvement of foreign language instruction in the United States. NSA has stepped into a gaping hole in the mission-agency coverage of an important national issue. There is nowhere else in the federal government where support for such research is available. By and large, the federal government spends all of its language research money in-house, leaving civilian foreign language instruction to its own devices.

In recent years, NSA has funded projects leading to the development of proficiency tests; the introduction of high technology into language teaching, including support for the creation of the principal professional organization in the field, the Computer Assisted Language Learning and Instruction Consortium

(CALICO); and the preparation of dictionaries in Ethiopian and Amharic and teaching materials in Georgian, Uzbek, German, Thai, Indonesian, Russian, Somali, Amharic, Polish, Spanish, Hebrew, Chinese, and Korean. It has even invested in telecommunications equipment for language instruction in the Hawaii school system. More recently, the Defense Language Institute, which provides language training for all military personnel, is creating a language research center that hopes to fund both civilian and academic research. The odd fact is that there seems to be relatively little general concern about DoD support—and from an intelligence agency, at that—of foreign language research. One exception was a proposed project on African languages. When it touched the African studies community at the University of California at Los Angeles, sparks flew.

Language and area studies

The golden age of amity between the military and academia during World War II extended to language and area studies as well. During the war, academics who were experts on other countries often found themselves brought directly into the war effort. For instance, the Office of Strategic Services, the forerunner of today's Central Intelligence Agency, recruited many academics, including the founding generation of many campus-based area programs. Many Japan specialists got their start by serving in MacArthur's occupation government in Japan.

It was the Army with its Specialized Training Programs that established programs on campuses across the country to train a corps of specialists on the languages and cultures of different societies. Some of these programs comprised an expansion of smaller programs already existing on a few campuses. Most of them were created from scratch. The purpose of these programs was to train intelligence officers or, in the case of occupation of those countries, military governors. Both the prototypical Specialized Training Programs and the faculty and students in those programs were the seedbed for the 600 or so language and area centers now spread throughout higher education. At the time, few if any scholars found this close tie with military affairs odd.

As an outgrowth of this period, the federal program that still provides the basic and most durable source of support for language and area studies was created, Title VI of the National Defense Education Act. This bill was enacted as a response to a military development, the launching of the Sputnik satellite by the Russians. Its first purpose was to create a cadre of civilian specialists sufficiently competent in the Russian language, and in a knowledge of Soviet society, to assure that in the future the United States would not again be caught unawares as it was in the case of Sputnik. The intelligence function was to be dispersed throughout academia.

While the initial impetus for Title VI of the National Defense Education Act grew out of the cold war, as the legislation took shape its coverage was extended to include all of the Third World. The Army Specialized Training Program model, this time to train civilians instead of soldiers, was re-created. The program was administrated in the Office of Education, rather than in a military agency, but the use of "Defense" in the title of the act made its rationale clear.[20]

---

20. "Defense" was later dropped from the title when this section was transferred into the Higher Education Act.

In subsequent years, the relationship between DoD and area studies followed the pattern of increasing stress that characterized DoD-university relations in general. American academics specializing in other parts of the world became increasingly alienated from DoD. This was in part because of a substantive issue, not a procedural one as in the case of the more general DoD-academic relations. Area specialists made little if any distinction between foreign and military policy, and they became more and more critical of the United States'foreign policy in general and its policy with respect to their part of the world in particular. Increasingly, sticky issues of foreign and military policy arose in Third World countries in which the professional interests of area specialists lie. The government as a whole, and the military in particular, rarely behaved the way academic specialists wished they would. Few academic Southeast Asia specialists supported the government's position in Vietnam. Few South Asianists supported American policy vis-à-vis Pakistan at the time of the separation of Bangladesh. Few Latin America specialists favor American policies with respect to Central America or the Caribbean. Few Middle East specialists support American policy in that area. To these grand issues can be added many smaller issues that occur in day-to-day foreign policy-making.

In general, area specialists, with their interest in a particular country or set of countries, tend to have a greater sympathy with the interests of the countries they study than do others in our society, including the military. This makes them difficult allies for the military, which sees its position as hard-headed realism viewed largely, if not entirely, from the perspective of American national interest. In fact, relatively few area specialists partici-

pate in the field of security studies at all, as the limited participation of area specialists in the Social Science Research Council's new Peace and Security fellowship program made abundantly clear. More generally, while political opinion within the area studies community is hardly monolithic, nevertheless the sense of alienation from American policy—in particular, military policy—is widespread and deeply felt.

From the perspective of academics, the most troublesome issues in DoD-academic relations reside in the intelligence section of the military, the part of DoD potentially most interested in the expertise of area specialists. In the eyes of many area specialists, DoD is identified with intelligence, and the boundary between DoD intelligence and other members of "the community"—in particular, the Central Intelligence Agency—have become blurred. As area specialists encountered American intelligence operations in the countries where they worked, and as the unsavory nature of some covert actions became a greater and greater subject for public debate, the desire to distance themselves from DoD increased.[21] This desire has been enhanced by the realization that academic colleagues in the countries where area specialists conduct their research will view rather harshly American academics thought to be identified with an intelligence service, perhaps going so far as to exclude them from research inside their countries. The ghost of Project Camelot lives on in Latin America and elsewhere. Indeed, scholars too closely identified with DoD or intelligence affairs are viewed by some of their American col-

21. For one of the many general examinations of this issue, see Sheldon Hackney, "Academia and Espionage," *Almanac* (University of Pennsylvania), 22 Sept. 1987, pp. 4-5.

leagues as volunteering for terminal leprosy.

This sense of disengagement from DoD by academic area specialists is accompanied within DoD by an increasing utilization of internal rather than external information sources for its policymaking. One reason for this trend is that many of the basic materials needed for analysis are too highly classified to share them with academics. Another is that the data used in academic analyses cannot be as current or as focused on immediate policy issues as is the huge amount of instantaneously available data within DoD. It is difficult enough for a DoD analyst to drink from the fire hose of information constantly piling on his or her desk from electronic eavesdropping, satellite surveillance, and daily digests of the content of foreign media, without adding tangential academic literature. It is true that there is periodic concern expressed at the highest administrative levels within DoD that this neglect of external opinion is costly. Top-level officials decry the myopia induced by DoD's exclusive concern with immediate information and the increasingly incestuous restrictions of data to internal sources. Nonetheless, the fact remains that the research product of the academic area studies community, with a few conspicuous exceptions, is relatively lightly utilized within DoD.[22]

By and large, over the years DoD has left the funding of language and area studies to National Defense Education Act Title VI, now the Higher Education Act Title VI, giving its moral support to appropriations when needed. In fact, on key occasions when other parts of the administration tried to curtail or abolish

22. For an extensive review of this situation, see *Defense Intelligence: Foreign Area/Language Needs and Academe* (Washington, DC: SRI International, 1983).

this program, DoD rallied to its support, including the sending of an almost unprecedented formal letter from one cabinet officer to another—from the secretary of defense to the secretary of education—protesting a proposed cut in the appropriation for Title VI. Currently, however, there is some indication that traditional DoD support for Title VI may be wavering.

In recent years, there have been several new experimental funding programs within DoD in support of campus-based language and area studies. The most successful was the creation of a several-million-dollar fund within the Office of the Secretary of Defense to support academic social science research on East European affairs. To administer this fund, an academia-administered National Council for Soviet and East European Research was established. After a few years of funding through the defense secretary's office budget, the responsibility for this program was transferred to the Department of State through an amendment of that department's authorizing legislation. One of the principal spokesmen for the congressional legislation—generally referred to as Title VIII—that made this transfer possible was a general in the Army, who happened to be a Soviet specialist. He later became the director of NSA. In subsequent years, many in the area studies field hoped that the State Department version of this model would be extended to other area studies groups; so far, it has not.

In addition to this program, focused specifically on support of Soviet and East European studies, DoD has introduced a number of experimental funding programs specifically aimed at providing support for academic area studies. Among these are the General Defense Intelligence Program, which joined together all DoD intelligence agencies in support of Third World language and area studies. The second is the Defense Advanced Language and Area Studies Program, which provides for advanced area studies by Defense personnel on American campuses. The third is the Defense Academic Research Support Program, which provides DIA with funds to contract for unclassified, publishable research on area studies topics of interest to it. By and large, these funds have been victims of the general cut in DoD's annual appropriation and the distribution of decision making concerning research support to the various services rather than the central planning section in the defense secretary's office. The fate of the Defense Academic Research Support Program is instructive for the present discussion.

One of the ways in which DIA proceeded to distribute funds under this program was to draw up a list of very broad topical domains in which it had a substantive interest and to invite area specialists to apply for funds to conduct research on topics of their choosing within these domains. As part of the publicity for this program, DIA sent a circular letter to each of the area center directors supported by Title VI, inviting research proposals from their faculty. This clear— and, in the eyes of many academics, dangerous—crossing of the DoD-academic line was the subject of a very heated debate within the area studies community. In the case of two of the world area studies communities, African and Middle East studies, their professional associations publicly decried such a widespread involvement of academics in intelligence matters and urged their members not to participate. In other world area studies groups, equivalent concerns were expressed.

The story of another DoD venture into potential funding for campus language and area studies is even more instructive. In 1982, the then secretary of defense, Caspar Weinberger, became personally concerned that the nation's campus-based advanced research and training enterprise was in danger of serious decline. Accordingly, the Army, on behalf of all of the services, was directed to contract for a major survey to determine the current state of affairs in language and area studies and to make recommendations for possible DoD support. Soon after, a working group was set up within the larger Defense-University Forum, a group of university presidents and high-ranking DoD officials founded for the express purpose of enhancing communication between DoD and the university community. The Association of American Universities was asked to conduct that study, and a year or so later, on the basis of an exhaustive survey, it submitted a report[23] with a list of 33 specific actions that DoD might take.

What followed within DoD is not unprecedented. While the report was greeted with general approval within DoD at the time of its appearance, making recommendations for specific DoD implementation was tasked—as they say in the military—to an unenthusiastic staffer and to an ad hoc academic-DoD committee. The report died. None of the recommendations has been implemented, with one exception. One of the proposals was for the creation of a congressionally mandated and funded National Foundation for International Studies to make available long-term funding for language

23. Richard D. Lambert et al., *Beyond Growth: The Next Stage in Language and Area Studies* (Washington, DC: Association of American Universities, 1984).

and area studies. To accomplish this purpose, DoD funded a project with the Association of American Universities to draw up a possible legislative charter for such a foundation. When the association's report proposing the charter was published, it drew a firestorm of hostile academic reaction. The principal objection was that the funding for the project had come from DoD. Both sides have agreed to let this particular project languish.

The point of this story is to dramatize in extreme form the problems in the current relations between DoD and the university community, and the international studies community in particular.

To a seasoned observer, it is quite understandable that stresses, disagreements, and misperceptions should have developed between DoD and the academic community. They are, barring another popular war, likely to continue and increase. From the perspective of the society as a whole, however, it is at least lamentable that the immense national resources that lie in our different institutional compartments cannot more effectively interact and reinforce each other. This is especially true in the area of international affairs, particularly in the determination of military policy upon which much of our country's future may depend.

The Military and Higher Education
in the USSR

By JULIAN COOPER

ABSTRACT: The military dimension of the Soviet higher-educational system is difficult to explore because of the striking scarcity of published evidence. There is a military presence at all universities and higher-educational institutes. It takes diverse forms, including special military departments and an elaborate system for the military-patriotic education of students. Research is undertaken for defense-related clients, although such research represents a modest share of the total military research and development effort. In general, the relationship with the military is harmonious, but this tradition of harmony, reflecting broader social attitudes toward the military in Soviet society, has come under strain in recent years as a result of changes in conscription policy. In the new conditions of *glasnost,* discontents have been able to find open expression and this may remain the pattern for the future.

Julian Cooper is lecturer in Soviet industry and technology at the Centre for Russian and East European Studies, University of Birmingham, United Kingdom. He is coauthor of The Technological Level of Soviet Industry *(1977),* Industrial Innovation in the Soviet Union *(1982), and* Technology and Soviet Economic Development *(1986). His current research is concerned with the contribution of the defense sector to the Gorbachev program for the revitalization of the civilian economy.*

THE military dimension of the Soviet higher-educational system is a topic that has received very little attention in the West. The reason is not difficult to establish: there is a striking paucity of evidence available in Soviet publications. In the recent period, however, two factors have served to modify the position: first, changes in conscription policy as it relates to students, and, second, the policy of *glasnost*. Problems of the relationship between the military and academia have now emerged into the light of public debate.

THE SOVIET HIGHER-EDUCATIONAL SYSTEM

Before considering the nature of the military involvement, it is necessary to establish briefly some of the basic features of the Soviet higher-educational system. The country has almost 900 civilian higher-educational establishments, or *vysshie uchebnie zavedeniia* (VUZy), as they are known in Russian, of which only 69 are universities. Of the remainder, some 65 are polytechnic institutes and 165 are institutes meeting the higher-educational requirements of particular branches of industry. Of the total student population of 5 million, almost 600,000, or 12 percent, study in the universities, the largest being those of Moscow, Leningrad, and Kiev. The armed forces have their own network of VUZy in the form of academies and higher military schools. The number of these has not been revealed in recent years but must be in excess of 150. Thus, of the total number of VUZy in the country, approximately 15 percent are military.

Some of the leading technical VUZy have long-established links with particular branches of the defense industry and meet most of their needs for qualified technical and managerial personnel. Notable examples are the Moscow Engineering-Physics Institute, which trains specialists for the nuclear industry, both civil and military; the Moscow Aviation Institute, many graduates of which work in the aerospace industry; the Leningrad Mechanical Institute, which has close links with the ground-forces equipment industry—graduates include the late Defense Minister Dmitrii Ustinov—and the Leningrad Shipbuilding Institute. The premier technical institute, the Moscow Bauman Higher Technical School (MVTU), appears to have close links with a number of branches of the military sector, including the missile space industry. An institute of increasing importance is the Moscow Institute of Electronics Technology at Zelenograd, co-located with the electronic industry's central research and development (R&D) facility. Some of these technical institutes are very large: the MVTU has more than 2300 teaching and research staff and some 25,000 students.

One distinguishing feature of the Soviet higher-educational system must be stressed from the outset: it accounts for a very modest proportion of the country's total research effort, and this has implications for the question of military involvement. Most fundamental research is undertaken at institutes of the USSR and republican Academies of Sciences, and a high proportion of the economy's applied research and development is performed at so-called branch R&D establishments affiliated with industrial ministries, military and civilian, and other state agencies. This leaves a modest share of total R&D expenditure, just over 5 percent, to the VUZy. Within the higher-educational system the universities and, in particular, a relatively small number of elite technical institutes account for the bulk of the

R&D undertaken; the universities alone undertake approximately one-quarter of the total. The leading universities are responsible for much of the basic research undertaken by the VUZy but, reflecting the academy's dominance in this field, such research represents only a minor component of the total VUZ R&D budget. In the mid-1980s, the overall proportions of VUZ R&D expenditure were 15:70:15 for fundamental research, applied research, and development, respectively.[1] More than 90 percent of VUZ R&D is undertaken for clients on a contract basis, the remainder being funded directly from state budget sources.

THE MILITARY RELATIONSHIP

What are the principal points of contact between the military and the Soviet higher-educational system? First, there is the research undertaken by civilian VUZy for the armed forces and the defense industry; second, the teaching role at the VUZy of specialists of the military sector; third, the existence at civilian VUZy of military-related faculties and departments; fourth, the arrangements for ensuring that students are imbued with what are considered to be the appropriate values toward the state and its armed forces; and, finally, the impact of the draft system on the higher-educational system. Consideration of these dimensions of the question cannot be divorced from the more general relationship between the military and society in the USSR.

The VUZy and military research

While the VUZy are responsible for a small proportion of the country's total

research effort, part of the R&D is undertaken on a contract basis with organizations of the defense industry and armed forces. Unfortunately, there is no information on the scale of this military component. It is likely that the military share is highest at those elite technical institutes oriented toward particular branches of the defense industry. It is known that such establishments as the MVTU and the Moscow Aviation Institute have long-term research links with enterprises of the military sector. In 1979, MVTU, for example, had such links with 300 industrial enterprises, civil and military, and had a total R&D budget of 20 million rubles.[2]

Contract research for military customers probably brings real benefits to the VUZy concerned. In circumstances in which even leading establishments have difficulties in obtaining advanced laboratory equipment and computers, close, stable research links with defense-industry clients are likely to facilitate access to scarce resources. In the Brezhnev years, the share of the state budget devoted to higher education suffered a steady decline and, as noted recently by Academician Yu. Ryzhov, rector of the Moscow Aviation Institute, this squeeze on resources has left even such high-prestige establishments as his own with a level of material provision well below accepted modern levels.[3] But, as Ryzhov has also indicated, industrial clients are not obliged to provide such material assistance, which means that VUZy cannot be sure of obtaining this form of support even from the most technologically advanced and wealthy military customers. This probably means that those VUZy without military clients

---

1. *Vestnik Vysshei Shkoly,* 1987, no. 6, p. 5.

2. *Vestnik Vysshei Shkoly,* 1980, no. 5, p. 46.
3. *Sovetskaya Rossiya,* 21 May 1988. See also Yu. Ryzhov, "Material' naya baza inzhenernogo vuza," *Kommunist,* 1986, no. 13, p. 59.

face even greater problems of supply. In these circumstances, it is not surprising that the level of computerization of the Soviet higher-educational system lags woefully behind U.S. or West European standards.[4]

Sovmestitel'stvo

It has long been a feature of the Soviet higher-educational system that some teaching is undertaken on a part-time basis by specialists from industry. Such outsiders may hold formal positions within the VUZy, including headships of faculties and departments. This practice of *sovmestitel'stvo* ("combined appointments") is one of the ways in which a defense-industry presence is established within universities and technical institutes. A notable example of recent years was the late general designer of missiles and launch vehicles, Academician V. N. Chelomei, who simultaneously headed a department at the Bauman Institute, the current rector of which, A. S. Eliseev, is a former astronaut.[5] Also typical was the late Academician Valerii Legasov, whose recent suicide was a tragic loss for Soviet nuclear science: he combined the post of first deputy director of the Kurchatov Institute of Atomic Energy with a chair at the Moscow Physico-Technical Institute and headship of a department of Moscow University.[6] This practice brings students into regular contact with leading representatives of military-related organizations and probably facilitates their recruit-

ment into the defense industry. There are additional benefits for the VUZy. Writing in 1980, a spokesman of the MVTU, which then had more than eighty representatives of industry teaching on a *sovmestitel'stvo* basis, noted that this outside participation eased the problem of obtaining laboratory equipment.[7]

Military faculties and departments

Some higher-educational establishments are well known for having military faculties geared to training officers with special skills. Examples include the Moscow Finance Institute, the Moscow State Conservatory, the Leningrad Institute of Physical Culture, and a number of leading medical schools. But many other VUZy possess military departments providing military training for reserve officers, and, as will be discussed, changes in the draft regulations have enhanced their importance during the 1980s.

Another aspect of the military presence at VUZy is civil defense. While there are grounds for doubting its practical efficacy, the Soviet Union has an elaborate formal infrastructure for civil defense. The universities and other VUZy maintain special departments for its organization and involve many of the teaching staff and students on a voluntary basis. Moscow University, for example, has a department of civil defense supplemented by volunteer faculty-level organizations, some headed by prominent members of the staff. Thus the chief of civil defense of one faculty, presumably that for philosophy, is R. Kosolapov, the well-known philosopher and former editor of the Party's theoretical journal, *Kommunist*.[8] Exercises are held on a regular basis,

4. Evidence is sparse, but an indication is provided by the fact that in the year 1982 the VUZy of the Ministry of Higher Education received just over 500 computers of all types. *Vestnik Vysshei Shkoly*, 1983, no. 6, p. 10.

5. *Podgotovka inzhenerov v Moskovskom vysshem tekhnicheskom uchilishche imeni N.E. Baumana* (Moscow: Vsshaya Shkola, 1983), p. 16.

6. *Pravda*, 30 Apr. 1988.

7. *Vestnik Vysshei Shkoly*, 1980, no. 12, p. 32.

8. *Voennye Znaniya*, 1988, no. 1, p. 31.

some taking the form of nuclear-attack simulations requiring students and staff to repair to the university's shelters. The civil-defense department undertakes research, including mathematical modeling of rescue procedures in the event of natural disasters or other emergencies.[9]

Military-patriotic education

It has long been a major concern of the Soviet state that the youths of the country should be imbued with values considered appropriate for the defense of the nation in the event of war. Military service is the rule for almost all young men, and there is particular concern that they enter the forces in a suitable physical and mental state. This applies to students in higher education as much as to other young people, and it involves a range of organizations and activities under the rubric of "military-patriotic education" (*voenno-patrioticheskoe vospitanie*). The principal bodies are the Communist youth organization, the Komsomol, and the mass paramilitary Voluntary Society for Assistance to the Army, Aviation and Navy (DOSAAF). In addition, university staff, especially those concerned with history and the social sciences, are expected to promote the appropriate values in the course of their work.

Each VUZ has its own branch of DOSAAF, to which, at least formally, many students will belong. It seeks to promote patriotic values and interest in military affairs, and also to develop the technical skills and physical fitness appropriate to military service. The latter are promoted through sports activities and a range of training programs enabling students to gain such skills as driving and vehicle maintenance, flying, parachute jumping, the use of radio equipment, and

marksmanship. DOSAAF keeps alive the memory of the Great Patriotic War through such activities as meetings of students with war veterans, visits to sites of battles, and the maintenance of war memorials. It also organizes ceremonies to mark significant military anniversaries.

Students and conscription

The Soviet Union has a system of conscription requiring all males over age 18 to undergo military training. Until the early 1980s, according to the 1967 Law on Universal Military Obligation regulating the draft system, students in full-time higher education could obtain deferments, postponing military service until after graduation. During their period of study, most students undertook some training in the military departments of their VUZy and then served for a reduced period compared with those entering at the normal age of 18. At the end of this foreshortened military service, the former students became reserve officers.

At the end of 1980, under the pressure of unfavorable demographic trends and, probably, also the Afghanistan war, this system was modified. According to the new regulations, which took effect in January 1982, deferments can only be obtained by students at higher-educational establishments considered of national importance. The list of such VUZy was to be approved by the USSR Council of Ministers on the basis of representations by the Ministry of Defense and the State Planning Committee.[10] This meant that the majority of students had to do their military service on reaching the age of 18, breaking their studies, if necessary, and returning to complete their degrees

9. *Voennye Znaniya*, 1987, no. 7, p. 82.

10. *Vedomosti Verkhovnogo Soveta SSSR,* 1980, no. 52, decrees no. 1121 and 1122, dated 17 Dec. 1980.

after the end of the normal term of military service, which is two years, three in the navy. As most students normally begin their higher education at the age of 17, it means in practice that they have to break their studies after their first year. Thus, with few exceptions—principally students with severe physical disabilities or exceptionally difficult family circumstances—military considerations impinge on all male students in the Soviet higher-educational system. While women can serve in the Soviet armed forces in auxiliary activities, they are exempt from the draft. But female students can, and do, join DOSAAF and participate in its sports and other activities.

The limited available evidence suggests that the 1980 restriction of deferments was enforced with ever greater severity, so that soon almost all students were forced to disrupt their education for military service. In November 1984 it was reported that first-year students of Moscow University had been conscripted after just starting their studies, and the university authorities acknowledged that this development required them to review arrangements for military-patriotic education.[11] If students even of Moscow University could no longer obtain deferments, it is likely that this privilege was retained by only a small number of elite technical VUZy, probably those most closely associated with the defense industry. One author, writing in the summer of 1987, noted that the number of VUZy retaining the right to deferments could be counted on one's fingers.[12] Conscripts at that time faced the prospect of active service in Afghanistan. It is not surprising that this policy provoked discontent and led to strains in the relationship between the military and higher education.

11. *Sovetskaya Rossiya,* 2 Nov. 1984.
12. *Komsomol'skaya Pravda,* 23 June 1987.

A TRADITION UNDER PRESSURE?

There is no doubt that the overwhelming majority of Soviet citizens, young and old, support the country's military effort. In the words of David Holloway, "By and large Soviet military policy appears to be regarded inside the Soviet Union as legitimate, and as pursuing legitimate goals."[13] This high degree of legitimacy of the military provides the general context for an understanding of the relationship between the military and higher education. The available evidence suggests that there is a strong tradition of harmony in the relationship. The military presence on campus, the research undertaken for military clients, and the teaching by defense-sector specialists are apparently regarded as perfectly normal and acceptable by most students and staff of the universities and higher-educational institutes. This does not mean, however, that there are no strains and conflicts. It is likely that over the years the implementation of particular aspects of the relationship has caused local difficulties. Under the Gorbachev leadership, in the new conditions of *glasnost* and pressure for democratization, an issue has emerged that has put the traditionally harmonious relationship under more general strain.

Gorbachev, glasnost, and the draft

The one issue that put real strain on the relationship between the military and higher education in recent years was the change in conscription policy adopted in 1980. This change took effect in 1982 at the very end of the Brezhnev period, but its real impact has been felt more recently

13. David Holloway, *The Soviet Union and the Arms Race* (New Haven, CT: Yale University Press, 1983), p. 162.

under Gorbachev. In the new conditions of *glasnost* it became possible to voice criticisms in public and to record manifestations of discontent. The military, hitherto shielded from public criticism, was required to undergo its own *perestroika*, involving revelations of inefficiency, corruption, mistreatment of recruits, and other negative phenomena. The military involvement in Afghanistan now began to receive more open acknowledgment, with public discussion of the often difficult problems faced by returnees and, in particular, the seriously injured. At the same time, the Gorbachev leadership was engaged in strenuous efforts to promote the cause of peace and disarmament, showing an unmistakable desire to reduce the Soviet military burden. These circumstances combined to make conscription deferment a sensitive issue of public debate from the summer of 1987.

It is reasonable to speculate that there was discontent with the new draft policy from its practical inception in 1982, but it was a debate in *Literaturnaya Gazeta,* the weekly paper popular with Soviet intellectuals, that finally brought the issue into the open. In the course of a roundtable discussion on the theme "Why do we have so few truly educated people?" the participants explicitly attacked the practice of forcing students to interrupt their studies in order to do military service. One of the most outspoken was Academician Boris Raushenbakh, the distinguished scientist, who noted that he had experience in teaching both students who came directly from school and those who had completed military service. "Those who come after the army," he declared, "are splendid executors, superbly organized, but one cannot obtain Newtons from them. Apparently, the creative abilities atrophy and deaden. As

a result within ten years or so we will have no one to do fundamental research, and this is dangerous for defense as well." Another participant, M. V. Vol'kenstein, corresponding member of the Academy of Sciences, concluded that "it is stupid and shortsighted to draft students into the army," while V. M. Mezhuev claimed that many students did not return after military service to complete their studies.[14]

This criticism of the new deferment policy provoked an immediate hostile response. Academician A. Fokin, a retired military specialist on chemical warfare, led the counterattack with a long list of prominent scientists who had served in the military, including Academy President Gurii Marchuk and the late missile space launcher pioneer Sergei Korolev. The significance of the inclusion of Korolev's name will not have escaped many of the readers of *Literaturnaya Gazeta:* Raushenbakh was for many years one of his deputies. Fokin also made the point that the country needed not just highly educated scientists, but patriotic specialists willing to serve their motherland.[15]

The principal response was in the pages of *Literaturnaya Gazeta* itself; the author was the Deputy Chief of the General Staff, Colonel General Makhmud Gareev. He began by suggesting that the roundtable participants were, in essence, attacking the entire system of universal military service and not just the lack of deferments for students. It was the duty of prominent scientists, cultural figures, and educators to promote correct attitudes toward the defense of the motherland. The implication was clear: Raushenbakh and his colleagues were not true patriots. An appeal to social justice was the core of Gareev's defense of the change

14. *Literaturnaya Gazeta,* 1987, no. 20, p. 11.
15. *Krasnaya Zvezda,* 16 May 1987.

of deferment policy: why should students be treated more favorably than the rest of the country's youths? The decision to allow the conscription of first- and second-year students had been made, he claimed, in response "to the countless requests from the Soviet public." The new policy had been successful. It had put an end to attempts by some immature young people to enter VUZy at all costs in order to avoid military service, and experience showed that more than 85 percent of students returned to complete their studies after military service. Like Fokin, he cited examples of eminent scientists who had served in the forces, including Marchuk and Korolev.[16]

The criticisms from Fokin and Gareev were echoed by other military spokesmen. Some linked the issue with what they perceived as an unhealthy growth of pacifist sentiments in Soviet society in general and among portions of youths in particular. These concerns came to a head in December 1987 at a meeting between Defense Minister Yazov and a group of well-known writers. Yazov, in an impassioned contribution, expressed doubts about the attitudes and maturity of some young people and clearly believed that intellectuals like Raushenbakh and some of the writers were, at least in part, responsible. "Some people do not want to serve," he declared. "It is hard, and he [the conscript] has not seen anything in his life, nor experienced anything. So what kind of young people are we bringing up?" Later, he said,

"A decision was made to call up tertiary education students. Look at the way *Literaturnaya Gazeta* burst forth: Oh my God, if one is to serve as a soldier, then he will never be a mathematician! He [Yazov pointed to one of the writers present] was a soldier and

became a writer. Many of you were first officers and soldiers, and then later became writers. 'He will not be a mathematician or scientist; he will not be this or that!'"[17]

This response of the military leadership indicates that the critics of the draft policy had touched a raw nerve, and one can only conclude that Raushenbakh and his colleagues had voiced a concern much more widely held in academic circles and possibly one also worrying sections of the political leadership. At the end of 1988, a brief statement in *Literaturnaya Gazeta* revealed that the critics had won a limited victory but one questionable on grounds of social justice: deferment was being reintroduced at a number of the most important VUZy. This presumably means that the group of privileged establishments was being widened, possibly in the hope of silencing the most articulate and best-connected potential critics, the students who attend the elite universities and institutes of the Soviet higher-educational system and their parents. Significantly, this announcement was not carried by any of the military newspapers or journals.

Is there any evidence of problems in the universities and institutes caused by the operation of the present conscription policy? For some time there has been acknowledgment of the practical problems of catering to students who have left their studies for two or three years and acknowledgment also of the new demands on military-patriotic education. Among the returnees are hardened Afghan war veterans, and there is evidence that some of these veterans experience difficulties in reintegrating into student life. At a number of VUZy, veterans are being

16. *Literaturnaya Gazeta,* 1987, no. 23, p. 11.

17. The proceedings of the meeting were shown on Soviet television in January 1988. See *BBC Summary of World Broadcasts,* SU/0053 B/1-4, 20 Jan. 1988.

drawn into the process of military-patriotic education, helping to prepare first-year students for the draft.

Noting that the discussion of military service and students was not based on sound evidence, a senior lecturer of the Gor'kii Agricultural Institute reported the results of a sociological survey conducted at the Institute. Of the students, 60 percent had served in the army or navy. The survey indicated that for the overwhelming majority the experience had enhanced the level of "civic maturity." Of those surveyed, 96 percent claimed that it had strengthened their intolerance of duplicity and deception, 70 percent that it had increased their opposition to alcohol abuse. Of those returning from military service, 72 percent claimed that they had resumed their studies without any difficulties. This leaves a sizable 28 percent who did experience problems, but the author did not elaborate. The thrust of the article, published, it should be noted, in the main military newspaper, is clear: there are no grounds for concern about the impact of the draft policy; on the contrary, military service for students helps to produce mature young citizens.[18] But the Gor'kii Agricultural Institute is a low-level VUZ; the findings may be untypical. That not all is well is shown by recent events at some institutions of higher status.

At Tartu University in Estonia, history students mutinied against military training in the University's military department, the incident being provoked by a lack of socks to go with their army boots! In this case, prompt action by the university authorities soon defused the situation. At Tomsk a more serious problem arose. Here students at the local university, the polytechnic institute, and an institute of radio engineering decided to boycott military training for reasons directly linked to the change in deferment policy. In this case, the authorities reacted more harshly, accusing the students of playing into the hands of imperialism. Problems arose because students returning from military service and attempting to resume their studies were being expected to spend a day per week in additional military training to gain officer rank, training that the students claimed was not adapted to their requirements as ex-conscripts. The standard of training was low, it ignored skills already acquired, and, in effect, amounted to little more than basic drill. At the university the military day was long, with no breaks for breakfast or lunch, and the training took place in a semiderelict building. The author of the article reporting the Tomsk situation, an *Izvestiya* special correspondent, gives the impression that the problems encountered at Tomsk are typical of many VUZy and, significantly, writing after the reported widening of the range of establishments covered by deferments, he declares without qualification that "now students have been deprived of the right to deferment from call up."[19] This suggests that any relaxation of the policy has been little more than cosmetic. The author, who applauds the student activism as a manifestation of the new spirit of *perestroika*, claims that the situation has now improved: the Ministry of Defense has reviewed and significantly reduced the program of military training within VUZy.

The Tartu and Tomsk student mutinies focus attention on particular sources of tension between the military and the VUZy, in particular the differential quality of staff in military and civilian depart-

18. *Krasnaya Zvezda*, 1 June 1988.

19. *Izvestiya*, 21 Apr. 1988.

ments and the limited degree to which the higher-educational authorities have control over their activities. It is apparent that at many VUZy there is a wide divergence between the academic levels of military and civilian departments; staff of the former rarely have higher degrees, and, indeed, some do not even have higher education.[20] In some VUZy, there are conflicts over staff numbers, with claims that the university authorities with the tacit support of the Ministry of Higher Education have been attempting to reduce the hours of teaching of the military departments as a prelude to reducing their staffs.[21] There have also been cases of civil-defense departments' returning falsified reports of their teaching, presumably to justify the maintenance of inflated staffs: in 1985-86 Kishinev Polytechnic Institute reported fictitious hours of teaching equivalent to two full-time staff appointments.[22] The position of the VUZy can be appreciated: they pay the salaries of the staff of military departments but have very little control over the latter's work.

The pacifists

The outburst by Yazov revealed that the military believes that sections of Soviet youths have become alienated from military values. Recent speeches and articles by representatives of the Main Political Directorate of the armed forces and DOSAAF confirm that this is now a matter of genuine concern. The chief of DOSAAF, Admiral G. M. Egorov, while acknowledging that the overwhelming majority of young people are, if necessary, "ready to go through fire and water in defense of our fatherland," has noted that there are cases of avoidance of military service, especially of service in the navy, which now extends to three years, compared with the two for other services. Egorov has not concealed his doubts about students in higher education. Noting that participation in military-technical types of sport was at a level inadequate to the needs of the country and the forces, he declared: "In the VUZy military-technical types of sport engage 2-3 percent of the students. The remainder sit and watch television or go to discos."[23]

There is no doubt that many military leaders believe that Soviet young people, especially students, are becoming soft and "unmanly," both physically and in terms of character. Military service, in the words of General Lizichev, chief of the Main Political Directorate of the Armed Forces, "brings out and strengthens the best qualities of the young person, develops feelings of citizenship, makes him more pure, stronger in spirit, internally rich."[24] "The army," another officer declared, "is a tough, men's school of life."[25] The Afghan veteran is often held up to Soviet youths as a model of the ideal Soviet male citizen—tough, responsible, pure in spirit, and fully committed to the goals and values of Soviet society.

Are young people in the Soviet Union, and students in particular, becoming more skeptical of military values and less willing to undergo military service? Are they now more likely to resent the military presence on campus so long taken for granted as a natural feature of the Soviet higher-

20. Ibid. To make matters worse, the author observes, in the military departments of the VUZy of Estonia there are almost no Estonian staff. It is likely that the same situation is found in other republics.

21. See *Voennye Znaniya,* 1988, no. 1, p. 35.

22. *Vestnik Vysshei Shkoly,* 1987, no. 6, p. 21.

23. *Kommunist Vooruzhennykh Sil,* 1988, no. 4, pp. 23-25.

24. *Kommunist Vooruzhennykh Sil,* 1988, no. 4, p. 15.

25. *Komsomol'skaya Pravda,* 13 June 1987.

educational system? In the more open society emerging under Gorbachev, it is becoming possible for young people to organize in so-called informal groups, that is, autonomous groupings outside the Party, Komsomol, DOSAAF, and other official public organizations. If students are critical of the military, they now have new opportunities to make their views known. Among the myriad informal groups now in existence—punks, rockers, *metallisty*, nostalgists, optimists, environmentalists, political discussion clubs, and so on—are the self-styled pacifists. According to Plaksii, a deputy director of the Komsomol's Higher School, they are usually older school pupils or first-year students at humanitarian higher-educational establishments. The pacifists are against war and opposed to military service. Unfortunately, there is no evidence on the scale of support for these pacifist views and whether it is such as to justify the military's concern. The more relaxed Plaksii regards the pacifists as naive and impressionable young people influenced by Western ideas and notes that similar pacifist views are held by some Soviet hippies.[26] Given that first-year students of the humanities form the core of the pacifists, it is likely that the movement is a response to the new conscription arrangements: such students, often children of members of the intelligentsia or Party and government officials, would previously have had no difficulty in obtaining deferments or avoiding national service altogether. The scale and significance of this movement should probably not be exaggerated, but its existence does indicate that traditional attitudes toward the military are not

26. See S. I. Plaksii, "Neformaly—Synov'ya ili pas'ynki?" *Politicheskoe Obrazovanie*, 1988, no. 7, p. 87.

without challenge in the present-day Soviet student milieu.

CONCLUSION

The Soviet Union is a country with a military that enjoys a high level of legitimacy. This is hardly surprising. The deep scars left by the Great Patriotic War and the sense of encirclement by powers of superior economic and technological strength have been the shared postwar experience of all Soviet citizens. The consistent pursuit of a policy of universal conscription means that most adult males have served in the armed forces and to that extent do not regard them as alien.

There is a well-established military presence at all Soviet universities and higher-educational institutes. The military relationship takes diverse forms. In comparisons with the United States, two features are of particular note: first, the extensive system of military-patriotic education designed to inculcate, not always successfully, attitudes and values supportive of the military and the Soviet state; and second, the fact that on the whole the VUZy are not major performers of R&D for the defense industry and, within the higher-educational system, the bulk of military-related research is undertaken by a few specialized technical institutes having close links with their clients. Students who enter these institutes must be very well aware of the military dimension and are likely to accept its legitimacy.

The available evidence, admittedly sparse, suggests that the relationship between the VUZy and the military has been harmonious with, until recently, few signs of strain. The single issue that has provoked real conflict has been the change of conscription policy in the early 1980s. Looking to the future, the position of the military is now less certain. Will a new generation of students entering the university under the conditions of *glasnost* and democratization, aware of the strong high-level emphasis on global cooperation and disarmament, accept so readily the traditional supportive values with respect to the military? The withdrawal from Afghanistan and demographic trends in the early 1990s may make possible a restoration of student draft deferments, thereby eliminating the principal source of strain. But at the same time, the Gorbachev leadership stresses social justice: why should students be privileged? The critics counter by stressing the country's need for talented specialists, an argument not without force at a time when economic modernization is at the top of the policy agenda. In my view, this consideration, coupled with concern that pacifist sentiments may gain greater sway among highly educated sections of Soviet youths, is likely to swing the balance in favor of the restoration of student draft deferments. New sources of strain may yet emerge, but, if they do, circumstances are now more favorable to genuine public debate: this openness must be considered one of the most impressive achievements of the Gorbachev leadership.

The View of the Big Performers

By STEVEN MULLER

ABSTRACT: The Applied Physics Laboratory (APL) of The Johns Hopkins University serves as an illustrative example of a large laboratory sponsored by a military service—the U.S. Navy—that is owned and operated by a major research university. After an examination of how APL functions within Johns Hopkins, the positive side of the relationship is examined. It is found primarily in the combination of national service and research and teaching collaboration with the University's academic divisions. The primary negatives are public controversy and the risks and burdens of the University's obligation. The position of The Johns Hopkins University has been and remains that classified research is not necessarily inconsistent with the purposes of the University and that a major public service is legitimately rendered by the contributions to national defense made by APL to the Navy, within limits set by the University.

Steven Muller, a former Rhodes Scholar, received his B.A. in political science from the University of California, Los Angeles, in 1948, a B. Litt. in politics from Oxford in 1951, and his Ph.D. in government from Cornell in 1956. He taught at Haverford College from 1955 to 1956 and at Cornell from 1956 to 1971. From 1961 to 1965 he served also as director of Cornell's Center for International Studies and from 1965 to 1971 as vice-president for public affairs. He came to The Johns Hopkins University in 1971 as provost and became president of the University in 1972.

THE military establishment of the United States contains within itself a substantial number of research laboratories of different sizes and with varied missions. In a few instances, however, laboratories exist that primarily do research for one of the military services or for another federal agency but that are neither owned nor operated by the federal government; instead, they are owned and/or operated by major research universities under government contract. These laboratories exhibit considerable continuity of mission and service, and the most obvious question raised by their existence is why the sponsoring military service or other agency finds it useful over time to rely on contracted service from a university-operated laboratory rather than an equivalent in-house facility.

WHY UNIVERSITY LABORATORIES?

For this reliance there appear to be several reasons. One of some significance involves staff compensation. As university employees, staff members of laboratories owned and operated by universities are not confined within government pay scales or bound by the civil service personnel system. Their compensation is subject to review by the sponsor in the course of contract negotiations, but it is essentially determined by the employing university. While such laboratory staff are not normally treated as university faculty and have no tenure, there is no question that their compensation is on the average substantially higher than that of staff employed in government-owned and -operated laboratories. It must also be assumed that the prestige and good name of the university involved helps to retain and attract personnel to university-owned laboratories and that some scien-

tists are more willing to work in university-affiliated laboratories than in government-owned or industrial facilities. This may be particularly the case when effective communication and some sharing of work exists between the sponsored laboratory and faculty and students in related areas of work within the university.

Perhaps the most persuasive reason for government sponsorship of laboratories owned by universities, however, is discovered in the relative independence from government control that is achieved by this arrangement. Such independence means not only that the government sponsor can look to the university to assure the quality of the laboratory's personnel and operations, but that the laboratory's work and status bear the hallmark of independence. University-owned laboratories are free to suggest research to the sponsor that does not originate within the sponsoring agency's own operations, and they are equally free to test and critique research desired by the sponsor concerning which they have reservations. The fact that an independent, university-based research effort is involved in the evolution of the sponsoring agency's work assumes considerable importance with regard to credibility and efficiency. Obviously, university-owned laboratories do not seek to offend their sponsors, and the sponsor must agree to fund the work to be undertaken, but even so, the independence of laboratory from sponsor remains significant. This is especially true when research conducted by university-owned laboratories leads to any kind of production. University laboratories may build prototypes but are not themselves engaged in production, which is purchased from commercial vendors by the sponsoring federal agency. University laboratories are, therefore, in a position to set specifications for production, assist the commercial producer in assuring quality control during the production process, and then test the finished product for meeting specifications. All this can be done effectively by the university laboratory as an independent resource employed by the government sponsor.

If these are at least some of the reasons why a military service or other federal agency finds it desirable to sponsor a major laboratory owned and operated by a university, the next question is why a university would wish to undertake such an obligation. Certainly, this is not done for profit, which is not permitted under the contracts that govern such arrangements. There is, in fact, only one reason why a university would operate a government-sponsored laboratory with a military mission, and that is simply to provide a service to the nation. The few large laboratories sponsored by the U.S. military establishment and owned and operated by universities in fact had their origins in earlier and simpler times, when the military needs of the nation were not only relatively uncontroversial but even paramount in public perception and when the notion that a university would voluntarily elect to perform this type of public service met with less cynicism—and criticism—than is the case today.

THE APPLIED PHYSICS LABORATORY

At this point, discussion will focus on the Applied Physics Laboratory (APL) of The Johns Hopkins University, simply because it is the one with which I am familiar and also because it illustrates most aspects of university-owned laboratories with military missions. APL is principally sponsored by the United States Navy, and its origins go back to World

War II. In 1942, there was an urgent need to perfect and develop the proximity fuze. A federal agency called the Office of Scientific Research and Development (OSRD) had been created to mobilize American science and technology in the war effort. Dr. Vannevar Bush, the OSRD chairman, had decided that a central laboratory for the development of the proximity fuze was needed. He called an old friend, Dr. Isaiah Bowman, then president of The Johns Hopkins University, who answered with eagerness to support national defense. The new laboratory was established under contract with OSRD and under management by Johns Hopkins on 10 March 1942.

The development and deployment of the proximity fuze was successfully and quickly completed, and a major beneficiary was the U.S. fleet. The Navy Bureau of Ordnance had become thoroughly familiar with APL, and in 1945 the Navy asked APL to assist the defense of the fleet by replacing antiaircraft shells with guided missiles. From this request a relationship between the Navy and APL was confirmed that has lasted to this day and that stipulates the defense of the fleet as APL's principal mission.

After the war ended, serious consideration was given by The Johns Hopkins University as to the suitability of continuing its management of APL. In 1947 an arrangement was made with an industrial organization, the Kellex Corporation of New York, to assume responsibility for engineering and product design and maintenance of staff and facilities, while overall research responsibility and management remained vested in Johns Hopkins. Predictably, this dual arrangement did not work well, and it lasted only a few months. The Navy, APL staff, and Johns Hopkins collectively decided that the preferable course was Johns Hopkins ownership and operation of APL for the indefinite future. On 26 March 1948, six years after the Laboratory had been formally established, APL became a full division of the University and has remained in that role ever since. Originally located in Silver Spring, Maryland, by the 1950s APL needed both new facilities and additional space. With funding supplied by the Navy, the University acquired land in Howard County, Maryland—roughly equidistant between Washington, D.C., and The Johns Hopkins Homewood campus in Baltimore—and the Laboratory moved onto these 365 acres. In recent years, APL has been operating with a professional staff of 1600, a total staff of 2800, and annual revenues and expenditures in excess of $300 million.

APL'S PLACE
WITHIN HOPKINS

Prior to a discussion of the advantages and disadvantages to The Johns Hopkins University of owning and operating APL under a Navy contract, a basic understanding is needed of just how APL functions within the University.

As has already been stated, APL has been established as a full division of the University for more than forty years, since 1948. The other divisions—among them arts and sciences, medicine, and engineering—are academic divisions headed administratively by deans. APL is a nonacademic division headed by a director. At Johns Hopkins, deans are appointed by the University president, customarily in consultation with the relevant faculty. The director of APL is also appointed by the University president, customarily in consultation with the outgoing director. Usually, the outgoing director will consult senior staff colleagues as well as representatives of the Navy

before making a recommendation to the University president, but there is no established process requiring such consultation, and the Navy has no formal voice in the governance or internal operations of the Laboratory. It is worth noting that there have only been five directors of APL—including the incumbent—since its founding in 1942 and that all after the first were selected from within the Laboratory.

Obviously, the major portion of APL's work is classified, and the APL campus therefore is the only campus of The Johns Hopkins University on which classified work is permitted. This does not mean, however, that the whole APL campus is closed. For decades, a major program in continuing studies has been carried on at the Laboratory, and outside visitors freely enter the classrooms and conference facilities, as well as the APL Library. Other facilities are restricted to visitors who must register prior to entry, and, as appropriate, entry is then further restricted to individuals with the required security clearances.

With respect to governance, APL's operations are under the supervision of the University president, just as is true of the academic divisions. The president and other senior members of the University Central Administration who deal regularly and directly with APL therefore have security clearance for this purpose. There is also a committee of the University Board of Trustees that supervises the work of the Laboratory on behalf of the Board of Trustees as a whole, and security clearance is required for the trustees who serve on this body. The Trustee Committee for the Applied Physics Laboratory customarily meets twice a year—spring and fall—for half a day, and senior representatives of the Navy, both civilian and military, are traditionally invited to attend these meetings, at least in part, as guests—they have no vote but are invited to participate fully in discussion. In addition, there is an Academic Advisory Board to APL, analogous to the regular academic councils or advisory boards of the academic divisions. The University president formally chairs each of the divisional academic councils or advisory boards, and also chairs the APL Advisory Board. In composition and mission, however, the APL Advisory Board differs from its regular academic counterparts. The Laboratory has no professors, so the APL membership of the Advisory Board consists essentially of the director, the assistant directors, major department heads, and other members elected by the Laboratory's principal staff. The non-APL members are the University president, the provost, and faculty representing the academic divisions that have the most interaction with the Laboratory—medicine, arts and sciences, and engineering. The APL Advisory Board also meets only twice a year—the academic councils or advisory boards of the academic divisions meet monthly or even more frequently during the academic year—and its main focus is on the non-defense-related APL programs that interact with University research and teaching in the academic divisions.

The Johns Hopkins University is uniquely decentralized geographically and administratively. There are three major campuses in Baltimore; the School of Advanced International Studies in Washington, D.C.; the Center for European Studies in Bologna, Italy; the Center for Italian and Renaissance Studies in Florence, Italy; the Hopkins-Nanjing Center for Chinese and American Studies in Nanjing, China; a number of Centers for Continuing Studies in Maryland; and APL. In this context, therefore, the high

degree of operational and fiscal autonomy of the Laboratory fits easily into the overall University system. No University dollars flow to the Laboratory. On the other hand, APL, as a full division of the University, participates in the formula funding in support of central university administration to which all divisions are subject on a proportional basis. The University also receives an annual management fee from the Navy for its services, which amounts to $150,000 per year, having been increased from the earlier annual amount of $75,000 during the period of high inflation in the 1970s.

A few other financial arrangements with respect to APL need to be mentioned in the interest of full candor and full understanding. The Laboratory itself receives a fee from the Navy as a fixed percentage component of the annual contract. Legally, this money is University money, because APL has no existence separate from being wholly part of The Johns Hopkins University. Indeed, APL's annual fee expenditures are subject to approval by the Board of Trustees. But the basic understanding between the Navy and the University provides that, except for the formula contributions to central administration mentioned earlier, fee income is to be spent on the Laboratory itself. Fee income is therefore a principal source of funding for the maintenance and enhancement of the APL physical plant. In addition, the Laboratory annually funds a small number of named research fellowships, to enable APL staff to teach and do research in one of the University's academic departments or to enable a University faculty member to conduct research at the Laboratory.

From the standpoint of the trustees, the administration, and the majority of those associated with The Johns Hopkins University, APL has been and continues to be a major asset of the institution. By means of APL, the University makes a major contribution to the quality, effectiveness, and survival capability of the United States Navy and thereby to the national defense and national interest of the United States. Most individuals connected with Johns Hopkins have almost no familiarity with the technical programs that make so great a contribution to the Navy, but those who are aware share the pride of the Laboratory itself in its significant accomplishments.

The quality and reputation of the Laboratory has been and continues to be consistently high. APL is an impressive place to visit, and contacts with members of the staff reveal an extremely high level of scientific and technical competence. At a research university such as Johns Hopkins, lower levels of ability would be readily recognized and would create unavoidable problems. No questions have been raised concerning the fact that APL's technical and scientific performance is at a level of quality fully commensurate with the University's highest standards.

APL'S CONTRIBUTIONS
TO THE UNIVERSITY

The most tangible ways in which APL contributes to the University are found in a growing array of collaborative programs with University faculty. The most enduring, varied, and rewarding of these collaborations have been in the area of biomedical programs, involving APL staff and faculty of the School of Engineering and the School of Medicine. These collaborations have involved cardiovascular research, ophthalmology, neurology, imaging, prosthetic devices, biophysics, and clinical engineering. In general, technology and skills originally applied to defense-related research have been suc-

cessfully applied to a whole series of biomedical problems and techniques. It is neither possible nor appropriate here to go into detail, but it is useful to note that nearly two dozen APL staff also hold appointments in the Schools of Engineering and Medicine, and an approximately equivalent number of members of the medical faculty also hold principal staff appointments at APL. Some hundred collaborators from APL, Engineering, and Medicine are working on some forty joint projects, over a hundred instruments have been developed for research and clinical applications, and hundreds of peer-reviewed scientific publications have appeared.

Another set of collaborations evolved from the development in the 1950s of APL's work into space, prompted by the Navy's need for satellite-assisted navigation and ocean mapping. APL's space department consequently developed links to the Department of Physics and Astronomy. The decision in the early 1980s to locate the Space Telescope Science Institute, the ground station for the Hubble Space Telescope, at The Johns Hopkins University was, of course, reached for many reasons, but unquestionably APL's record of space research was among them. APL has been involved in a number of non-defense-related space missions and built much of the instrumentation designed and required for the Hopkins Ultra-Violet Telescope, a project of the University's Department of Physics and Astronomy funded by the National Aeronautics and Space Administration.

Substantial joint work also goes on between APL staff and members of the faculty of the School of Engineering. The largest collaboration, however, involves APL's participation in the continuing-education programs of the Engineering School. At Johns Hopkins, the bulk of continuing education is carried on at the graduate level, and one of the largest centers of continuing education in engineering is located at APL, with annual enrollments of over 2000 students. Formally, this is an activity of the School of Engineering, but the majority of those serving as faculty for the programs located at APL are APL staff, who thus have the opportunity to teach. On the one hand, APL's location away from the Homewood campus headquarters of the Engineering School is a major advantage, enabling the educational program at APL to attract a geographic population substantially removed from the Baltimore area. On the other hand, a microwave relay network links several of the Johns Hopkins campuses, and two interactive computerized classrooms—sharing simultaneous sound and sight—make it possible for a single instructor to teach a single class in two remote settings—in this case, at APL and Homewood.

A variety of other interdisciplinary APL-University faculty interactions exist in other fields as well, including applied mathematics, environmental protection, oceanography, and transportation studies. Over the last two decades, the volume of these interactions has continued to increase, and the participation of APL in the rest of the University, and vice versa, has become ever more substantial. While the single Navy contract remains both the core and the bulk of the Laboratory's annual funding, non-defense-related smaller contracts or grants from agencies such as the National Aeronautics and Space Administration, the Department of Energy, the National Institutes of Health, and departments of the state of Maryland now represent somewhere between 10 and 18 percent of APL's annual work load. Overall, then, APL is in these many—and a few lesser—ways a major

asset of The Johns Hopkins University. In the circles that are most closely familiar with its work, it clearly contributes favorably to the University's reputation. Also, it has now been part of Johns Hopkins for decades and is therefore accepted as part of the University in a far easier manner than would be the case were it of only recent origin. For alumni and friends of Johns Hopkins who have an interest in national security, or who at least respect the University's contribution to national defense, APL is a showpiece and an attraction.

THE NEGATIVE SIDE

The other side of that coin is that APL is also a source of controversy within Johns Hopkins, among faculty, alumni, and friends, and in the public arena. At the most thoughtful level, there continues to be debate as to whether a university committed to freedom of teaching and research can and should legitimately own and operate a laboratory most of whose work requires secrecy.

At another level, APL is, of course, perceived as part of the nation's military-industrial complex, and its work on the defense of the nuclear subsurface as well as the surface fleet—in addition to its work on the guidance systems of missiles capable of bearing nuclear warheads—links it directly to the nuclear threat. At this level, the principal argument is that University operation of APL is an immoral act, life threatening and evil especially insofar as it contributes directly to the possibility of nuclear warfare. Antinuclear activists have made the Laboratory a regular target of protest over many years, both at APL's campus and at other Johns Hopkins campuses and in public settings. When protest of this kind involves trespass and/or damage to property, as happens with some frequency, the resultant arrests and court proceedings receive wide publicity and the University is pilloried. At present, some of the University's adherents of Physicians for Social Responsibility and a more aggressive organization called the Committee for the Conversion—of APL to peaceful purposes—have played leading roles in public debate and protests directed against the Laboratory and the University's responsibility for its work.

Obviously, the most intense controversy surrounding APL arose during the period of the Vietnam war, particularly in the late 1960s and continuing into the early 1970s. This was the time when several other universities that had been operating defense-related laboratories under contract disengaged from these operations. While there was vigorous protest at Johns Hopkins as well, the trustees and administration reaffirmed the University's intention to continue the APL mission and ownership and operation of the Laboratory.

More recently, APL accepted significant specific research tasks requested by the Strategic Defense Initiative Office and has been publicly identified as a significant Strategic Defense Initiative contractor. Predictably, the decision to engage in this work—which involves fundamental scientific and technical problems of great interest, but whose details are classified and therefore cannot be made public at this time—represented an additional factor in the sporadically recurrent protests concerning the Laboratory.

On the one hand, opposition to APL has consistently remained at a relatively low level and cannot to date be characterized as a major problem of the University. On the other hand, it cannot be dismissed as trivial. While there is literally no evidence that members of the faculty or students have either selected Johns Hopkins or have failed to select or have separated from the University because of APL, it is clear that a minority of alumni and other potential donors are withholding support because of their opposition to the Laboratory. Over the long run, a great deal of time and energy on the part of University Central Administration as well as on the part of the APL leadership is devoted to responding to queries about and challenges to the Laboratory's presence at Johns Hopkins and its work. The University President's Office has been vandalized at least once as part of a protest against APL, demonstrators have appeared at private residences off campus as well as on campus, and critical comment continues to appear occasionally in the public information media.

There are some other negatives for the University with respect to APL that are not publicly controversial. A huge responsibility and significant risk attach to the ownership and operation of such a large and complex laboratory plant and staff. APL's funding under the contract is annual, which means that legally there are no guarantees as to its future. A sudden decision by the Navy to terminate its relationship with APL would present an enormous problem. As a hedge against such an unforeseen but not impossible development, the contract has allowed for the establishment of a multi-million-dollar stabilization and contingency reserve, whose accumulated capital mostly serves as a working capital reserve for the Laboratory but would, of course, be fully replenished were the contract to be terminated. Even so, however, the disestablishment of APL would not be easy, and the conversion of all of the facilities and persons involved to other purposes, absent the Navy contract, does not appear to be possible.

The contract itself presents all of the difficulties associated with the need for annual renewal and refunding. Negotiations with the Navy at both the uniformed and civilian levels of administration, and concern with congressional attitudes and initiatives, have become virtually a year-round preoccupation, not only of APL leadership but of relevant personnel in University Central Administration as well. There are also all of the normal problems of audits, personnel administration, litigation, and so on. Above all, however, is the University's full responsibility for the scope, nature, and quality of the Laboratory's work.

THE UNIVERSITY'S POSITION

The subject of the University's responsibility prompts some conclusions and reflections on APL as part of The Johns Hopkins University that may or may not apply at other universities with analogous laboratories but that do appear to be valid for this particular case. The Johns Hopkins position with respect to APL is unambiguous. Operation of the Laboratory is a national service that remains a valid mission of the University as long as the mission is clear and the service remains necessary; as long as the quality of service performed remains at the highest possible levels of scientific and technical excellence; as long as the University's authority over the Laboratory remains absolute; and as long as the means are furnished to do the job to be done with the needed resources.

Secret work is not acceptable as part of the University's academic divisions. On a separate campus, however, it is no less acceptable than the long-established acceptability of professors' obtaining clearance to participate in classified activities—or than the proprietary secrecy of

commercial corporations—away from the academic campus. The military needs of the nation are as much a part of the national interest as non-defense-related concerns. University contributions to national defense are traditional in times of need and can range all the way from the operation of Reserve Officers Training Corps programs to the ownership and operation of laboratories.

Not all military research needs, however, are equal in terms of suitability for university involvement. Aside from the question of whether a university can make a unique and irreplaceable contribution by operating a laboratory, the university must also be able to choose and limit the work to be done. Defense-related work should not conflict with the university's academic programs. There was, for instance, a case, in years past and not at Johns Hopkins, in which a university laboratory under military contract covertly offered counterinsurgency training to nationals of certain Southeast Asian countries, to the embarrassment of the academic community involved in that university's program of Southeast Asian studies. A university will regard certain kinds of defense-related research as unacceptable. Johns Hopkins, for example, will not involve itself in the development of chemical or biological weaponry. A university also has the responsibility to facilitate and encourage the availability of scientific and technological innovation developed as part of defense-related research to non-defense-related science and technology.

APL's mission remains focused on the defense of surface and subsurface vessels of the United States Navy. Not only is this clear as an overall context, but the specific activities of the Laboratory are carefully defined and reviewed within

APL and by University administration and trustees on a continuing basis. APL has been asked but has refused—and would not be permitted by the University to accept—wholly black contracts, that is, commitments to conduct work knowledge of which would be limited only to those directly involved and would exclude officers of the University with responsibility for the Laboratory. The degree of comfort with APL at The Johns Hopkins University rests above all on the fact that the nature of the work done is fully known to the University's leadership, which therefore can and does take responsibility on an informed basis.

Johns Hopkins has been fortunate that APL has had directorship of exceptional competence throughout its existence and that the quality and morale of the Laboratory's professional staff also has been consistently very high. The degree of integration of APL into the University and the expanding interaction between Laboratory staff and University faculty represent another boon, among other things reinforcing faculty respect for APL's level of effort and performance. The basic contract, with all the occasional difficulties attached to it, has nevertheless proven to be an effective and—relatively!—simple administrative mechanism. And, obviously, what will soon be five decades of successful and productive existence represents an experience capable of generating momentum and respect of its own.

Were there no such history and no APL, would The Johns Hopkins University today agree de novo to undertake responsibility for an analogous effort in support of the national defense? The answer is, of course, doubtful, but if the University could be assured that its services were urgently needed and that the experience would be as positive as the APL experience has been at Johns Hopkins, the response very likely would be affirmative.

ANNALS, *AAPSS*, **502**, March 1989

The Good of It
and Its Problems

By RICHARD D. DeLAUER

ABSTRACT: The evolution of the present relationships between the military establishment and the nation's universities in the performance of defense-related research and development is summarized. The present relationships and organizational concepts have their roots in the years immediately prior to World War II. The 25 years from the end of the war to 1970 were particularly significant in shaping today's military-university collaboration and cooperation. The Vietnam era and the high-inflation period of the 1970s created many problems for the established relationships; particularly troublesome were the antiwar sentiment on many campuses and the decision by certain universities to reject classified military research efforts. Renewed support of science and technology endeavors in the late 1970s brought about certain improvements in the military-academic environment; innovative initiatives in research and development administration in the 1980s have further improved these relationships. Representative problems and solutions and possible future directions of these very important relationships are reviewed and explored.

Richard D. DeLauer served as under secretary of defense for research and engineering from 1981 to 1984. Formerly an executive vice-president and director of TRW Inc., he has held various executive positions in the field of defense and space systems acquisition and integration management. A 1940 Stanford University graduate, he received a Ph.D. in aeronautics and mathematics from Cal Tech in 1953. He is now chairman of the Orion Group Ltd., an aerospace consulting organization.

FEDERAL and state support of basic and applied research activities at the nation's colleges and universities has a long history, going back well into the last century. In the Reconstruction era following the end of the Civil War, limited governmental support in the form of sponsored academic research projects began in a small way as a result of the growing social awareness and philosophy of providing governmental support to our nation's institutions of higher learning.

THE PERIOD 1938-46

It was not until the years just prior to World War II, however, that the military services—the War and Navy departments—began specific programs of underwriting relevant scientific and technological research projects at a number of the leading colleges and universities in the country. Research into various aspects of science and technology related to military aviation was given primary emphasis by the forerunner of the U.S. Air Force—then the Army Air Corps, as a part of the U.S. Army—as well as by the Marine Corps, which then, as now, was a part of the Navy Department and under the technical leadership of the Navy in scientific and technological endeavors.

With the approach of World War II, the need to accelerate the application of the emerging technologies of the day to military capabilities in such fields as radio communications, radar, sonar, and high-speed, long-range aircraft required the services of the finest research and engineering talent that could be found. A rich source of such talent was the laboratories and research facilities of many of the scientific and engineering universities of the nation.

The list is extensive, but some of the more notable facilities at that time were the Radiation Laboratory associated with the Massachusetts Institute of Technology (MIT), the Guggenheim Aeronautical Laboratory at the California Institute of Technology, and the nuclear physics research laboratories at the University of California, the University of Chicago, and Columbia University. In many such academic facilities, resident faculty and graduate students were pursuing basic research in fundamental aspects of advanced scientific and technological areas. Such effort made significant contributions to the military capability of the United States as it initiated its rapid buildup at the start of World War II.

Due to the obvious success of the military-university relationship during World War II, it was not surprising that the often ad hoc arrangements established during the 1938-46 time period provided the basis for the adoption of more formalized relationships between the military establishment and the academic community in the postwar period. It was during this period that the roots of our present university-military relationships were formalized and strengthened and the extensive interactions between the various components of the Department of Defense (DoD) and the academic community assumed the basic characteristics that they retain to the present day.

THE PERIOD 1946-70

It was during the 25 years after the conclusion of World War II that a significant number of very important university-military relationships and organizational concepts evolved. Our university student bodies and faculties were swelled by returning veterans. Many of these students were interested in pursuing scientific, engineering, and other technical fields of study, and a great many faculty members had direct knowledge and experience of military operations and organizations.

In significant respects, the postwar academic community was in favor of supporting the national defense and was comfortable with the concept of university-military cooperation in research and development endeavors. This environment was conducive to the establishment and refinement of the military-university relationships that continue to exist today. The postwar period shaped the manner in which a very significant portion of our military research and engineering is carried out at the present time.

Two major research laboratories were operated under contract by the University of California for the Atomic Energy Commission and are still so operated by the university for the commission's successor, the Department of Energy: the Los Alamos Scientific Laboratory and the Lawrence Livermore Laboratory. The Los Alamos Laboratory, at Los Alamos, New Mexico, was originally established as an element of World War II's Manhattan Project, while the Livermore Lab was established in 1952 by splitting it off from the Lawrence Radiation Laboratory on the University of California campus in Berkeley. Their major missions involve support of the development of nuclear weapons; however, they are now involved to a great extent in military research activities well beyond their original research and development roles in the field of atomic weaponry. At present, both laboratories are engaged in important research efforts related to the Strategic Defense Initiative program as well as various other advanced military projects.

Similarly, two other World War II laboratories became institutionalized in the university-military cooperative research infrastructure during the late 1940s. These are the Jet Propulsion Laboratory at Cal Tech and the Lincoln Laboratories of MIT. The Jet Propulsion Laboratory is an outgrowth of the advanced aeronautical and rocket-propulsion work done under the direction of Dr. Theodore Von Karman at the Guggenheim Aeronautical Laboratory at Cal Tech, and Lincoln Labs evolved from the MIT Radiation Laboratory.

It was during this postwar period that a novel organizational concept evolved as a result of the need to have a more technically competent and stable scientific work force than could be realized either with technically trained military officers or with skilled civil servants operating under existing pay-scale and work-force ceilings. This need was satisfied by creation of the Federally Funded Research and Development Center (FFRDC) concept. FFRDCs are established as not-for-profit organizations, typically with their own governing board of trustees. They provide scientific, technical, and program support to the various military service elements on a contract basis and thus function as continually available, as-needed technical resources.

At the present time, most FFRDCs are not directly affiliated with a particular university, although in the 1950s and 1960s many of them established advisory or steering-committee relationships with a particular university. For example, the Center for Naval Analyses has such an advisory relationship with the University of Rochester; the Charles Stark Draper Laboratory to MIT; the Naval Underwater Laboratory to Pennsylvania State University; and the recently established Software Institute to Carnegie-Mellon University.

In addition to such on-going FFRDC steering-committee relationships, many university faculty members and administrations also play an important role in the definition and direction of military research and development through involvement in the activities of various DoD scientific and technical advisory bodies. Such organizations as the Defense Science Board, the Army Science Board, the Air Force Scientific Advisory Board, and the Naval Research Advisory Committee all have, on a continuing or rotating basis, a significant number of university faculty people as members or senior advisers.

It was also during the 1946-70 period that significant funding of directed on-campus defense-related research projects was instituted. With the creation of the Defense Advanced Research Projects Agency in the late 1950s and the enhancement of the existing research management organizations of the individual military service branches—in particular the Office of Naval Research and the Air Force Office of Scientific Research—many significant university research programs were instituted in direct response to specific military requirements. The research efforts sponsored by the Defense Advanced Research Projects Agency have been especially impressive. Computer time-sharing, packet switching, and the fifth-generation computer project are particularly noteworthy of such advanced research efforts carried out by academic organizations.

The DoD has been the sponsor of major computer development and computing application programs since World War II. DoD support of the research on the electronic numerical integrator and calculator and of Dr. John Von Neumann's work at the Princeton Institute for Advanced Study are but two examples of their interest and foresight. Elsewhere in this volume, detailed reference is made by Dr. Leo Young to the antecedents and

present status of DoD efforts in electronics and computing.

It is of interest to note that the fifth-generation computer project was the result of research carried out in the late 1960s under sponsorship by the Defense Advanced Research Projects Agency of projects at MIT, Stanford University, and Carnegie-Mellon University. The outstanding results of this advanced research were not exploited by the U.S. computer industry but in the late 1970s were appropriated by the Japanese and have provided the foundations for the rapidly expanding international computer market of the 1980s.

It is my opinion that the failure of the U.S. computing industry to exploit the fifth-generation effort was primarily the result of complacency on the part of the large mainframe producers. In the late 1960s and early 1970s, they had a virtual monopoly on the computing and information-system market. It was not until the Japanese had entered the field in a major way that the industry finally took notice of what was happening.

THE PERIOD 1970-85

During the early 1970s, growing anti-war sentiment on many of our prestigious university campuses in reaction to the Vietnam conflict, in conjunction with a policy of benign neglect of the scientific and academic community by the Nixon administration, brought about significant negative changes in the military-university relationship.

Also sometime in the early 1970s, the leadership of the National Academies of Engineering and Science, in particular the National Academy of Science, embarked on a program of disengagement from the DoD. For several years, it had been the practice of the Defense Science Board to conduct its annual summer studies at the National Academy of Science's study facilities at Woods Hole, Massachusetts. These summer studies involved conceptual problem definition and the description of advanced research programs needed to meet projected or perceived military mission needs.

Suddenly, in 1973 or early 1974, the Defense Science Board was informed that the Woods Hole facilities would no longer be available for subsequent summer studies and that meeting accommodations should be sought elsewhere. This rift was caused in large part by deep-rooted political differences between certain influential individuals in the National Academy of Science and the Nixon White House. Also, the Vietnam conflict created sharp schisms in the DoD-academic community. Even the Reserve Officers' Training Corps programs came under attack, and a number of universities withdrew their sponsorship of these long-established military training programs. The net result was that the DoD and much of academia went their separate ways for most of the last half of the 1970s.

At about the same time, strong objections to classified military-related research projects started to be heard from some faculty members on certain university campuses. Within the next few years, many of the nation's center-of-excellence universities adopted policies that precluded any further DoD-funded classified research on the campus under university management.

These universities did not in general prohibit specific faculty members from engaging in classified military research efforts as individuals, but they objected to further university involvement in such classified projects. The adoption of such a policy by MIT was the primary reason for the Draper Laboratory to disengage itself from MIT management to become an independent FFRDC. It also led to the Jet Propulsion Laboratory's becoming primarily a National Aeronautics and Space Administration support center rather than continuing to obtain some degree of support from classified military projects.

This separation between the universities and the military continued throughout the 1970s and was further aggravated by the growing neglect of the defense technical base by the two Nixon administrations. In the lexicon of the defense budgeters, all "tech-base" funding is contained in the 6.1, 6.2, and 6.3A budget categories in the Defense Appropriation Act legislation passed by the Congress for each fiscal year. During the years from 1968 to 1974, these accounts were substantially reduced, thereby causing the DoD to make severe reductions in the funding of research and exploratory development in all quarters: government laboratories, industrial contractors, FFRDCs, and the universities.

This negative trend in the funding of the technology base was a consequence of the Nixon administration's lack of regard for support of applied science and basic research. Also, it was due in part to the continued shortsightedness of the individual services in cutting their tech-base funding first whenever their overall funding levels were reduced, as was the case during the early 1970s.

It was not until the advent of the Ford administration, principally as a result of the leadership of Vice-President Nelson Rockefeller, that military research and technology resumed a position of importance in influencing White House budgetary deliberations, and the precipitous decline in 6.1 and 6.2 funding was reversed.

During the Carter administration and with new leadership in the Pentagon, technology-base funding continued to make positive gains in relative terms, but the double-digit inflation of the late 1970s resulted in there being less money to spend in real terms. As a consequence of this, as well as the anti-classified-research stance of many universities, the brunt of the reductions in research funding during these years was largely borne by the universities.

Also during the Carter years, other changes were taking place at many of our preeminent technical universities. Campus unrest and antimilitary attitudes were subsiding, new and aggressive leaders were taking charge of university administrations, and the continuing growth in inflation was causing severe fiscal problems at nearly all of the nation's academic institutions. During this period, increasing numbers of students were applying for admission to institutions of higher learning, but the universities were finding it increasingly difficult to expand their facilities and equipment to accommodate the growing demand.

The economic pressures of the time, such as the high rate of inflation, resulted in significant increases in the cost of living and major increases in the costs of higher education. This inflation also created many opportunities for engineering and scientific students to enter the work force at very attractive starting salaries upon completion of their four years of undergraduate study. These salaries were particularly attractive when compared with the minimal compensation the universities could afford to pay their graduate teaching assistants. Thus many students chose to forgo their plans for graduate degrees, and this placed an even greater burden on the universities due to a shortage of graduate students, at the same time

reducing the nation's pool of advanced-degree scientists and engineers.

All of these influences had a multi-faceted effect on the university-military relationship. Doctoral candidates have traditionally supplemented the faculty by providing teaching assistance to undergraduates and at the same time supplying much of the lower-cost labor that is needed to carry out the research programs supervised by the senior faculty members. Consequently, a marked change in attitude toward outside-sponsored research programs occurred both in the ranks of academia and in the university administrations. In the early 1980s, many of the universities began actively to seek ways and means of obtaining increased levels of federal research support, even if this meant accepting classified DoD projects.

THE 1980s

With the election of Ronald Reagan to the presidency and the rapid increases in defense funding, it soon became evident that although science and technology were recognized national imperatives, large increases in funding from other than U.S government departments were not going to occur in the form of research grants or nonspecific funding of scientific endeavors. Although the National Science Foundation had become more active and influential in the administration from a policy point of view, it did not receive a large infusion of funds for the support of basic and applied research programs, when compared to the increases received by the DoD. Also, the National Science Foundation's subsequent increases in funding fell far short of those seen in some of the departmental budgets.

Thus it was left to the functional organizations such as Health and Human Services, the Department of Energy, the

Department of Commerce, the National Aeronautics and Space Administration, the Environmental Protection Agency, the Department of Transportation, and the DoD to fund and support any increases in university research activities. At the same time, new leadership was being installed in both the National Academy of Science and the National Academy of Engineering as well as at the National Science Foundation.

As a result, a more collaborative and collective attitude toward university research emerged. One of the initial programs was that of recognizing the urgency for supporting a major upgrading of university instrumentation capabilities. This upgrading was needed in terms of both quality and quantity. The research advances being made in all fields—biology, medicine, and others—resulted in an exploding need for sensing, computing, and analytical capabilities. Coupled with the neglect of the 1970s in this area, the scientific instrumentation needs of the early and middle 1980s were tremendous. While this program received broad interagency support, the DoD, was the major monetary contributor. Still, the funding that could be made available was far from adequate to meet the need.

During this same period, the DoD recognized that it was imperative that old differences be resolved and a new agenda for the university-military relationship be established. As a result, the DoD-University Forum was established, cochaired by the under secretary of defense for research and engineering and a selected university president. Although this forum did not include representatives of every university in the nation, it did have the participation of nearly all major institutions that had a demonstrated or expressed interest in being involved in military research, in-

cluding most of the recognized centers of excellence throughout the country.

In addition, the forum was endorsed by many of the nation's higher-education organizations, such as the American Council on Education, the Association of American Universities, and the National Association of State Universities and Land-Grant Colleges. The forum met quarterly for more than three years and was reasonably successful in creating an improved atmosphere of trust. Additionally, it helped to reconfirm the traditional mutual interdependence of the military establishment and the university community in research endeavors.

The members of the forum included university presidents and the senior leadership of the DoD research and technology organizations. Particular issues were considered when specific clarification of policies or positions was needed and firm recommendations were explored and adopted. The long-term success of this university-military forum depended to a very great extent upon the willingness and commitment of the individual military services and defense agencies to provide the necessary levels of funding. Unfortunately, the accomplishment of this goal was a continuing and only partially successful battle.[1]

Although the military benefits of a strong research infrastructure within the nation's universities were obvious to many knowledgeable parties both within the DoD and elsewhere, the secretary of defense, the secretaries of the Army, Navy, and Air Force, and the chiefs of

1. A complete account of the creation, organization, and deliberation of the forum is contained in *Report of the DoD-University Forum* (Washington, DC: Department of Defense, Office of the Under Secretary of Defense for Research and Engineering, Dec. 1984).

staff of the uniformed services did not always share this understanding, particularly if it meant trading force structure and modernization for research and exploratory development. Keeping the technology base adequately funded and then assuring that the universities received their appropriate share of support was a yearly exercise of major proportions that far exceeded the actual amount of money being expended.

In order to provide additional sources of funding for university research activities, an initiative was undertaken by the under secretary of defense for research and engineering to induce the defense industry to share some of its discretionary independent research and development funds in a collaborative way with the university research organizations. Since the level of reimbursement received by an individual company for its independent research and development program depended upon an annual assessment, or grade, that it received from a tri-service technical review, the grading criteria were modified to include an explicit credit to a company's overall independent research and development program evaluation if the program included an explicit amount of effort placed with university research organizations. Simple in concept, this initiative also was of only limited success, again as a result of the constraints exercised by the uniformed service personnel who made up the review and grading teams.

Notwithstanding these bureaucratic impediments, the university-military relationship has markedly improved in the past few years. With the low inflation rates of recent years, the available levels of funding have also increased in real terms compared to what was available during the 1970s.

One of the significant facts that emerged from the deliberations of the DoD-University Forum was that not all universities had become averse to the subject of classified military research. It was, in general, the most prestigious institutions—those with large private endowments, and those who were extremely selective in matters of faculty selection and student qualifications—that were the most adamantly opposed to conducting classified research projects during this period of time. As it turned out, a majority of the universities represented on the forum were—and are— receptive to selected classified activities, provided they are adequately funded for the effort.

This is not to suggest that the DoD has adopted a policy that only classified research should be carried out or that all military research being performed by universities should be classified. A great deal of past and present military technology has its roots in basic and applied research efforts that initially had no obvious military application and thus no original military classification. It is frequently only when specific military applications or the use of the results of such general research to meet specific military missions becomes involved that the matter of classification enters the picture.

This issue of early classification was the subject of a major debate and policy determination within the Reagan administration in 1983 and 1984. Due to ideology and ambition, certain DoD officials, while unfamiliar with the short half-life of basic research, were determined to control restrictively the dissemination of all DoD technical information. Their technique was to classify everything. They took the position that, notwithstanding any agreements or assumptions at the beginning of an activity, they and they alone could, after the fact, classify the effort and thereby place restrictions on the publication and dissemination of the results of the research.

At the same time, those DoD officials who had responsibility for the maintenance of the defense technology base through the conduct of the overall defense research and development program had a completely opposite view. Their position was that DoD-funded basic research conducted at the universities should be unclassified without exception. If a project or activity appeared to require the protection or restrictions of a military classification, the issue should be addressed at the outset, thereby allowing the universities either to accept or to reject the assignment for that particular effort.

This difference of opinion was debated within the administration for nearly a year. Finally, with support from the Office of Science and Technology in the White House, the views of the research and development people in the Pentagon prevailed, and President Reagan issued a policy directive that provided that military research conducted by universities should be unclassified unless specifically excepted at the initiation of the effort. This policy also permitted the unrestricted dissemination of the results of such research in symposia and publications both domestically and internationally. If this policy had been resolved in the other direction, all the previous years of work to foster a collaborative environment between the military and the universities would have been undone.

If the military-university research relationship is to remain strong and productive, each community must fully understand the limits and constraints of the other. There will always arise, from time to time, issues that, if overstated or permitted to become contentious, could place undue strains on the collaborative relationship that is critical to the success of the effort.

The DoD must never lose sight of the fact that a healthy, productive, and independent system of higher education is an absolute imperative for the continuing security of the nation. Conversely, the academic community must constantly maintain its understanding of the fact that the structure, attitudes, and political constraints of the military establishment must be accepted and adhered to. If the collective leadership of both communities fails to maintain sufficient maturity and resiliency to maximize the strengths and minimize the differences between the two, then more heat than light will be the result.

The DoD has never claimed to provide research support to all segments of the academic community in a comprehensive or evenhanded fashion. It must be selective, and at times even arbitrary, in distributing its support. This derives from the reality of the facts about both missions and resources. The services see their mission generally as to provide matériel and personnel to create a certain force level. If their resources are limited, as they always are and will be, then they must be allocated selectively and a set of priorities must be established. Consequently, there will always be haves and have-nots among various university faculty groups.

What is needed is a recognition of this dilemma by all governmental sponsors of research—both state and federal—as well as by private industry sources, so that continuing efforts are made on an overall basis to attempt to achieve the most evenhanded result possible. Such an integration has not been carried out very effectively in the past fifty years, but it seems evident that the situation has been improved to some extent in recent times. Continued and even greater efforts in the future to further minimize this perceived inequity will help the overall military-university research relationship a great deal.

THE CONTINUING NEED
AND FUTURE PROBLEMS

From a policy point of view, there is no question on the part of any responsible military or civilian leader in the DoD that basic and applied research must be funded adequately and carried out effectively. History is replete with examples of how significant advances in technology can alter and have altered national security strategy and the balance of world power.

Therefore it is obvious that military-related research must be conducted to assure that we have taken into account such revolutionary eventualities. The question has always been one of what kind of research and how much, not of whether to conduct it. As has been discussed herein, this question results, for the most part, in differing answers that depend largely upon the leadership that is involved. This leadership is not only in the DoD and the administration, but also in the Congress and in the national perception of the security risk involved.

What is not recognized universally is that, from a policy viewpoint, a strong military-university relationship provides the pool of educated people that continually is required for strong professional military and industrial leadership. A healthy research effort provides, both to the services and to industry, the trained and knowledgeable scientists and engineers upon which both depend. It contributes to a comprehension and under-

standing of mutual problems as well as mutual opportunities.

Notwithstanding all such positive aspects, the DoD-university research relationship continually is being questioned and harassed. From critics within the DoD, it is a question of content and degree. From those outside, it is always a question of why particular research activities are being supported or what possible use to the military there can be in some pinpointed area of research. The DoD is still required to operate under the Mansfield Amendment of the 1970s, which insists upon a litmus test of direct military relevancy before the research can be funded and undertaken. Sheer stupidity!

What, then, will the future bring? Only some educated speculation is possible, but for now, the following scenario can be forecast:

1. The DoD budget will not grow appreciably in the next 10 years.

2. About 6 percent of the gross national product is what can be expected for the DoD.

3. Unless drastic revisions are made in what force structure is really needed to provide effectively for our national defense, the tech base will continue to suffer.

4. Further neglect of the tech base will be counterproductive, because what is needed is the capability and resources to make the military establishment less and less labor intensive.

5. If some form of national service can be instituted, coupled with education as an incentive, a major shift in how DoD resources are allocated will result.

6. Such initiatives as these will require strong and enlightened civilian leadership. The military services will not initiate such changes on their own.

7. Without a strong and affordable military establishment, the United States will decline as a world power as well as a global military force.

8. We must have a continuing, imaginative, and healthy research and development program if we wish to continue to enjoy a strong national security posture, within both our industrial and our military establishments.

ANNALS, *AAPSS*, 502, March 1989

Military Funding
of University Research

By VERA KISTIAKOWSKY

ABSTRACT: Since the mission of the Department of Defense (DoD) and the purposes of the universities do not coincide, the question whether this is pragmatically important is examined. Relevant events in recent history are mentioned, and the reasons given in support of DoD funding of university research are summarized. Five consequences of such support are then discussed: distortion of the balance between research fields, change of emphasis within research fields, classification and other restrictions, consequences for graduate students, and political consequences. Social responsibility for the end use of research is then considered, and it is suggested that social responsibility should become an important university criterion of excellence. The article concludes that the negative consequences of DoD funding far outweigh its perceived benefits, and suggests that the universities should work to establish a strong civilian base of research funding instead of lobbying for DoD support.

Vera Kistiakowsky is professor of physics at the Massachusetts Institute of Technology. She received her A.B. from Mount Holyoke College in 1948 and her Ph.D. from the University of California at Berkeley in 1952. She has done experimental research in nuclear and particle physics and in astrophysics. She has spoken and written on arms-race topics. She is on the board of the Council for a Livable World and cochair of United Campuses to Prevent Nuclear War.

THE increases in the funding of university research by the Department of Defense (DoD) since 1979[1] have rekindled the debate, important in the 1960s, about the appropriateness of military funding of university research. Theoretically, this question may be addressed by comparing the mission of military research and development (R&D) with the purpose of the universities. The mandate of DoD is to provide for national defense, and it funds R&D that is pertinent to that purpose both directly, within weapons programs, and indirectly, by the support of basic and applied research relevant to perceived present and future defense needs.[2] Furthermore, DoD is interested in the production of an adequate supply of scientists and engineers in the fields important to anticipated future weapons research, development, testing, and evaluation (RDT&E) programs.

The purpose of the universities was originally purely educational, and today the emphasis on the education of students is still central. However, their role as institutions involved in the increase of

knowledge and understanding through independent scholarship and research is now at least equal in importance. Consequently, in an ideal university, academicians would be free to choose research topics on the basis of their intrinsic interest and importance. Rapid research progress depends upon the open discussion of results and free exchange of ideas, and these are also vital to the quality of graduate education, since much of this is carried out through research. The breadth and excellence of the research opportunities available to graduate students therefore have a direct connection with the quality of the next generation of scholars.

Thus, theoretically, the mission orientation of DoD support and the possibility of security restrictions inherent in work related to national defense do not coincide with the purposes of the universities, based as they are on freedom and diversity. However, the possibility remains that this is pragmatically unimportant. This article addresses the pragmatic question by first commenting briefly on some relevant recent history. This is followed by a summary of reasons commonly given in support of DoD funding of university R&D and then by a discussion of the undesirable consequences. In the final sections, some ethical issues are raised, and the question whether there should be funding of university R&D by DoD is considered.

---

1. Between fiscal years 1979 and 1986 there was a 48 percent increase in the fraction of federal funding of university research and development that came from DoD. In this period, the fraction of federal funding of all research and development from DoD grew from 49 percent to 69 percent. U.S. Department of Defense, *Report on the University Role in Defense Research and Development* (hereafter referred to as *URDRD*), prepared for U.S. Congress, Committees on Appropriations, Apr. 1987, tab. 4, p. 16.

2. "In particular, the DoD award process must take into account considerations of both technical quality and relevance to well-defined missions of the Services." U.S. Department of Defense, *The Merit Review Process for Competitive Selection of University Research Projects* (hereafter referred to as *DDRMRP*), prepared for U.S. Congress, Committees on Appropriations, Apr. 1987, p. 6.

DoD FUNDING
OF UNIVERSITY RESEARCH
SINCE WORLD WAR II

To understand current attitudes concerning DoD support of university research, it is useful to review the antecedents of the present situation. Prior to World War II there was little funding of university research by the federal government. When the United States began to be drawn into that conflict, there was an urgent national need for expertise and service in a number of disciplines. The traditional functions of the universities became secondary as educational facilities were used for officer training and faculty took leave to work on war-related research and in advisory capacities. Laboratories for research on weapons and defense problems were established under the aegis of the universities.

When the war ended, the successes of radar, the Norden bomb sight, and the atomic bomb convinced the military of the desirability of continuing the link with the universities. In 1946 Congress set up the Office of Naval Research (ONR) with the explicit purpose of supporting academic basic research that might someday be of military use to the country, and graduate-student education to insure a source of scientists who could be called on in a future crisis. There was, concurrently, strong support for the establishment of a civilian agency which would have oversight of all basic research, including that of interest to the military, but it took eight years before various problems were resolved, and, when the National Science Foundation began operation in 1953, it was for nonmilitary research only. In the meantime, the Korean War had shifted the emphasis of ONR to research more directly relevant to military problems, and the Army Research Office (1951) and the Air Force Office of Scientific Research (1952) were set up explicitly to do mission-oriented research. The Mansfield Amendment to the 1970 fiscal year (FY1970) DoD Authorization Act, which stated, "None of the funds authorized to be appropriated by this Act may be used to carry out any research project or study unless such project or study has a direct and apparent relationship to a specific military function or operation,"[3] made explicit what had in fact been the case since the Korean War. However, there is a memory of that early period of benign support of basic research by ONR—particularly among those who were graduate students in that period— which is still recalled in discussions of the current situation.

DoD funding of university research continued to rise until it reached a maximum in 1964-66.[4] The subsequent decrease was the consequence of the Vietnam war, both due to budgetary constraints and because the campus opposition to the war led to an erosion of the university-DoD relationship. Both the university administrations and the DoD did not at first appreciate how serious and widespread was the criticism of this increasingly unpopular war, and DoD exacerbated the situation by shifting support from basic research to applied research and, in some cases, even to classified development work in direct support of weapons systems. This fueled further opposition on the campuses involved, and lessened the acceptability of DoD funding. In the same period, both Presidents Johnson and Nixon cut back on the funding of RDT&E programs among other efforts to control the defense budgets while continuing to escalate the war.

The decreases in DoD university funding continued until 1975.[5] This was a

3. Stanton A. Glantz and Norm V. Albers, "Department of Defense R&D in the University," *Science,* 22 Nov. 1974, p. 706.

4. The maximum in current dollars was $0.295 billion, in 1966; in constant (FY1987) dollars, $1.063 billion, in 1964; and in percentage of RDT&E funding to academic institutions, 4.5 percent, in 1965. *URDRD,* tab. 1, p. 7.

5. The minimum in current dollars was $0.197 billion, in 1974; in constant (FY1987) dollars, $0.451 billion, in FY1974; and in percentage of RDT&E funding, 2.4 percent, in FY1974 and FY1975. Ibid.

period marked by a general lack of interest in science on the part of the federal government, most visibly evident in the abolishment of the Office of the Presidential Science Adviser, and the decreases in DoD support were not offset by increases through other agencies. Various new programs supporting R&D in areas related to needs of the civilian economy were meagerly funded, and, for the first time since the depression, substantial numbers of graduating doctoral scientists and engineers experienced difficulty in finding employment in the fields in which they had been trained. As a consequence of these employment prospects and the decreased graduate-student support, the numbers of doctorates granted in mathematics, the physical sciences, and engineering began a decline in 1971 which continued into the 1980s.[6]

The middle 1970s were marked by an increasing return to influence of those who believed in military solutions to the U.S.-U.S.S.R. competition. Concurrently, rapid advances in technology suggested the possibilities of new generations of advanced weapons systems. The need for R&D to support these concepts and the predictions of a future shortage of scientists and engineers led to a revived interest on the part of DoD in university research. Universities, pinched for funding, responded positively. The Association of American Universities and individual university presidents testified to Congress, urging support of increased university funding by DoD. This resulted not only in an expansion of the number of awards in traditional single-discipline projects

but also in the more structured University Research Instrumentation Program in FY1983 and the University Research Initiative in FY1986.[7]

In 1983 President Reagan's Star Wars speech expressed the ultimate belief in a technical solution to the threat of nuclear war. This led to the establishment of the Strategic Defense Initiative (SDI) in 1984 and rapidly increasing funding for a narrow range of RDT&E programs.[8] The president called for a scientific crusade to make his dream a reality, but instead there was mounting criticism of the program by the scientific and technical community, criticism which contributed eventually to increasing congressional resistance to the rapidly escalating presidential budget requests for SDI. In what was publicly acknowledged as a move to obtain the implicit support of university scientists and engineers for the program, the Strategic Defense Initiative Office of Innovative Science and Technology (SDIO/IST) was established to fund university research in relevant areas. The political nature of this initiative resulted in widespread criticism, and more than 3700 senior research personnel signed a pledge not to request or accept funding from SDIO/IST.[9] An approximately equivalent number of academics did accept the funding quietly, but attempts to

7. *DDRMRP*, p. 5.

8. In FY1984 existing RDT&E programs with a combined budget of $0.9 billion were taken from other branches of DoD and placed under the authority of SDI. The budgets for successive years were $1.4 billion in FY1985, $2.8 billion in FY1986, $3.2 billion in FY1987, and $3.6 billion in FY1988. Since FY1987 Congress has approved substantially less than the amount requested in the president's budget. "The Strategic Initiative" (Briefing paper, Union of Concerned Scientists, June 1987); "News & Comment," *Science,* 26 Feb. 1988, p. 967.

9. The signatories were faculty at 110 colleges and universities that received 75 percent of all academic physical science and engineering funding

6. The number of doctorates awarded in mathematics, the physical sciences, and engineering reached a maximum of 9237 in FY1971 and then decreased to 6696 in 1981. Peter D. Syverson, *Summary Report 1981: Doctorate Recipients from United States Universities* (Washington, DC: National Academy Press, 1982), p. 6.

rally substantial and explicit support for the program were unsuccessful.[10]

REASONS FOR DoD FUNDING OF UNIVERSITY RESEARCH

In any discussion of DoD funding of university research, a number of arguments are commonly given in favor of this support. These include reasons why DoD funding is good for the universities and why university participation is good for DoD and national security.

The DoD belief that university funding is in its own best interest stems from the fact that weapons increasingly depend on the latest advances in technology. Basic and applied research are both necessary to maintain the existing U.S. superiority in the relevant fields, and the university connection supplies high-quality research at very low cost. This is enhanced because of the participation of graduate students and recent postdoctoral research workers with fresh ideas and state-of-the-art knowledge and techniques. It also ensures that there will be a future supply of scientists and engineers trained in fields relevant to weapons work and that these will include some of the next generation of scientific leaders. Finally, it maintains a nucleus of personnel at the universities with some familiarity with national

defense problems on whom DoD can call in a crisis, and creates a group of scientists and engineers who are more likely to be politically supportive of DoD RDT&E.

The central reason why many have considered DoD support to be good for the universities in recent years is that succinctly encapsulated in the Sutton principle,[11] although this is usually couched in more diplomatic language concerning the current availability of funding. Additional reasons are also given, one of which is that association with DoD RDT&E programs gives access to forefront technological and scientific knowledge not available elsewhere. Using this, scientists can lay the basic research foundation, and engineers can help construct solutions to problems at the cutting edge of the nation's capability, a process from which the universities cannot afford to be excluded. In addition, DoD funding makes available valuable state-of-the-art equipment and facilities that could not otherwise be acquired. Furthermore, it is argued that the DoD is a superior supplier and overseer of contracts and grants, with fewer restrictions and reporting requirements, better able to make long-term commitments than the civilian agencies.

The proponents also argue that DoD is a benign source of funding for basic research and that the content of the research it supports is independent of the funding source. The argument continues that security considerations have not been a problem in the past and points out that industrial support of research involves proprietary constraints on publication and patent rights, arguing that the suitability of DoD funding should be evaluated on a comparative basis, not in

by DoD and the National Science Foundation in 1985. Lisbeth Gronlund et al., "A Status Report on the Boycott of Star Wars Research by Academic Scientists and Engineers" (May 1986), p. 6.

10. A statement in support of SDI by the Science and Engineering Committee for a Secure World was placed in the *Congressional Record* by the House Defense Subcommittee. There were 80 members listed, of whom 46 were possibly senior academic scientists and engineers. One was a Massachusetts Institute of Technology graduate student, incorrectly listed as "Research Assistant, High Energy Physics, Fermi National Laboratory." "Testimony of Dr. Martin J. Hoffert on SDI," *Congressional Record,* 15 May 1986.

11. Willie Sutton, renowned as a bank robber, when asked why he robbed banks, replied, "That's where the money is."

isolation. Finally, the proponents conclude, even if DoD involves special constraints, it is currently such a small percentage of university support[12] that the effects would be unimportant.

UNIVERSITY RESPONSIBILITY FOR PUBLIC SERVICE

Another reason given for university participation in DoD-sponsored research is that this is in the national best interest and thus obligatory. The university indeed has changed from its origins as a purely scholarly institution, to one which has public service as one of its several functions. The Massachusetts Institute of Technology (MIT) makes its service commitment explicit:

Service: As a modern university and social institution, MIT recognizes an inherent obligation to serve its students and alumni, the professions, the world of scholarship, and society. As part of this obligation, the Institute seeks to serve the community and nation directly through its faculty and through the use of its facilities and administrative resources whenever there is a compelling need to which it can respond without impairing its primary functions.[13]

Consequently, if a convincing argument can be made that a DoD component of university R&D is in the interest of national security, then this is an obligation that should be honored, unless there are serious consequences for education and scholarship. There are those who would argue that university participation in DoD RDT&E is important to national security. For most people, the current unpersuasiveness of this statement is due

12. In FY1986, DoD supplied 16.4 percent of basic, 12.0 percent of basic and applied, and 16.4 percent of all research support from the federal government. *URDRD*, tab. 4, p. 16.

13. *Policies and Procedures* (Cambridge: Massachusetts Institute of Technology, Sept. 1985), p. 17.

to disagreement with military buildup as an effective road to national security, and even greater lack of agreement with DoD on what is desirable in terms of weapons programs.

Another argument usually given to support the service aspect is that the results of the R&D have broad nonmilitary applications and are thus good for the civilian economy. This rests on past examples of spin-offs, notably semiconductors and computers in the 1950s and 1960s, but there is considerable evidence to show that the increasingly narrow weapons focus of DoD R&D has had decreasing yield for nonmilitary purposes. Finally, it is claimed that inclusion of university scientists and engineers in the DoD programs will result in a broader and more balanced point of view, beneficial to the programs. This claim is hard to prove or refute, due to the lack of comparable situations with and without civilian leavening. It must also be weighed against the other outcome, which has been experienced, that the participants will become uncritical supporters of the programs that fund them.

DISTORTION OF THE BALANCE BETWEEN RESEARCH FIELDS

Turning to the reasons for opposing DoD funding of university research, the first that will be addressed is the effect of such mission-oriented support on the balance between research fields. Perhaps the best-known example is nonmilitary, the effect of the foundation of the National Cancer Institute on the field of biology, but in the physical sciences the establishment of the Materials Research Laboratories by the Defense Advanced Research Projects Agency in the late 1960s resulted in a similar disproportionate growth. The argument that the percentage of DoD funding is still small—16.4 percent of federal funding of university research in FY1986[14]—overlooks the fact that it is concentrated in a limited number of fields and therefore can account for a substantial increment in their funding. The SDIO/IST was initially very outspoken about its purpose with respect to research balance and stated the following as one of its objectives: "Mount a mission oriented, basic research program that drives the cutting edge of the nation's science and engineering effort in a direction that supports existing SDI technological development thrusts and points the way for future new initiatives."[15]

A direct consequence of such shifts in university research funding is a change in the availability of graduate-student financial support. This is reflected in the relative number of doctoral degrees granted in the various fields and, therefore, in the future field distribution of research personnel nationally. For example, cryptology and related mathematical fields are now supported generously by the National Security Agency, and consequently these are areas offering increased graduate-student support relative to other fields of mathematics. This is mission-oriented funding and produces professionals in that field for national defense purposes, not because of the needs of academic science, the fundamental importance of the research, or its relevance to other national problems.

CHANGE IN EMPHASIS
WITHIN RESEARCH FIELDS

Not only is the balance between fields affected, but the nature of the research within the fields is changed, although this point is vigorously denied by most participants in such funding. However, this is not borne out by the following DoD description of the assistance that it offers in proposal preparation:

Interested researchers typically will [contact] . . . DoD program managers. . . . These interactions will help potential proposers to decide whether their ideas coincide with DoD research needs. In instances where ideas do not initially "fit" DoD programs, the potential proposer may, with information provided by a research administrator, modify his or her approach to accommodate DoD needs.[16]

An early 1970s study of DoD-funded research at Stanford came to the following conclusions:

We found that individual scientists paid with DoD money did indeed view themselves as being involved in objective searches for truth and that they did not consider their searches to be intimately connected with the immediate military problems in Indochina. . . . Our study demonstrated that the military had developed a rational, well-administered program to define research priorities in terms of current and projected military needs and to purchase R&D from universities based on these needs.[17]

The change in emphasis within fields, like the shift in balance between fields, should be considered in terms of its consequences. This is a difficult problem because one cannot know the course that scientific inquiry would follow if it were free from constraints. Both balance and emphasis are shaped by policy decisions on what will be funded, either by the federal government or by other sectors. However, few academicians would support the suggestion that the criteria of fundamental interest and scientific excellence should be overshadowed in the

14. *URDRD*, tab. 4, p. 16.

15. U.S. Department of Defense, *The Strategic Defense Initiative Organization/Innovative Science and Technology Office*, Mar. 1985, p. 3.

16. *DDRMRP*, p. 5.

17. Glantz and Albers, "Department of Defense R&D in the University," p. 706.

areas of basic research by perceived military needs for information and research personnel.

In applied research, emphasis has an even more evident relation to the question of what is best for the nation. The preferential funding of weapons programs has had its economic consequences, witnessed by our deteriorating civilian industrial infrastructure, our negative balance of trade, and our mounting deficit. We have not invested in civilian research at a level adequate to these pressing problems, and will continue to suffer the consequences until we do. While there have been spin-offs from military R&D in the past, the yield has been low compared to what would result from research directed at the civilian problems. As Simon Ramo, former president of TRW, said, "In the past thirty years, had the total dollars we spent on military R&D been expended instead on those areas of science and technology promising the most economic progress, we would probably be today where we are going to find ourselves in the year 2000."[18] Furthermore, the money-is-no-object, technology-limited approach to the problems of weapons development is in sharp contrast to the cost considerations important for competitiveness in civilian markets, limiting transfer possibilities.

The differences between military and civilian RDT&E also have their consequences for the scientific and technical work force. Scientists and engineers trained in DoD areas and attitudes cannot be casually transferred to civilian problems. This is nationally important since an increasingly large fraction of the scientific and technical work force is engaged in military RDT&E,[19] and thus it becomes increasingly difficult to find personnel suited to solving problems related to the civilian economy.

CLASSIFICATION AND OTHER RESTRICTIONS

Another issue is whether the security considerations inherent in military R&D are compatible with an academic setting. There is not agreement among the universities on the issue of acceptance of classified research. The Georgia Institute of Technology and Carnegie-Mellon University have units carrying out classified work on their main campuses, while MIT and Johns Hopkins University have laboratories for classified work which are geographically separate. The decision in 1984 by the California Institute of Technology (Cal Tech) to establish the Arroyo Center, an off-campus think tank for classified army problems, led to a heated debate between the faculty and the administration, followed by a reversal of that decision. The concluding event embodies the most prevalent attitude, which is that the requirements of excellence in education and research at universities are incompatible with classification. This was one of the strong concerns voiced about SDIO/IST-funded research, since it is in a category—6.3, advanced development—which is usually classified. In response to this concern, Dr. James A. Ionson, director of SDIO/IST, released a memo which states:

The conduct and reporting of research performed on university campuses when sponsored by the IST Office of the SDIO, although funded out of budget category 6.3, will be treated as "fundamental research." ...

18. Simon Ramo, *America's Technological Slip* (New York: John Wiley, 1980).

19. Of all scientists and engineers, 228,506, or 13 percent, were employed in national defense work in 1982. U.S. National Science Foundation, *The 1982 Postcensal Survey of Scientists and Engineers,* 1984, tabs. B-12 and B-27.

However, when there is likelihood of disclosing operational capabilities and performance characteristics of planned or developing military systems, or technologies unique or critical to defense programs, the contract will stipulate that the responsibility for the release of information resulting from IST research belongs to the sponsoring office.[20]

Shortly thereafter, the White House released National Security Decision Directive 189, which states that "to the maximum extent possible, the products of fundamental research [at the universities will] remain unrestricted . . . except as provided in applicable U.S. Statutes."[21] Here "fundamental" includes basic and applied research, or categories 6.1 and 6.2. Note that in both cases there is language retaining the right to restrict.

There are other research-related restrictions possible under the Arms Export Control Act of 1976, the Export Administration Act of 1985, and a provision in the Defense Authorization Act of 1984 giving the secretary of defense authority to limit access to information subject to the export-control laws. One of these is the denial of participation in research and of access to facilities by non-U.S. citizens, which is potentially worrisome at the universities where an increasing fraction of graduate students are foreign. Furthermore, it is an embarrassment if distinguished foreign scientific visitors must be excluded from portions of the campus.

The limitation of access applies not only to the research but also to public presentations of the results. There was a crisis at the April 1985 conference of the Society of Photo-Optical Instrumentation Engineers, where a last-minute decision

by DoD meant that closed sessions had to be hastily arranged for 20 percent of the papers.[22] Twelve of the major scientific and technical societies subsequently co-signed a letter to Defense Secretary Caspar Weinberger stating that they would not allow closed sessions.[23] DoD responded by saying that it would sponsor separate closed sessions as adjuncts to the meetings of those societies.[24] Because of the inclusiveness of the lists of what may be restricted, there have been recurring problems of this nature. Agreements are reached on one situation only to have a new application of restrictions occur in a different area.

A related concern arises from the linkage of unclassified university research with classified research elsewhere to which the university workers require access. This leads to the necessity for security clearances, even for graduate students, and consequently to a limitation on their ability to discuss technical issues underlying their work. The extent to which this occurs nationally is not known, but it is not negligible in universities with strong DoD ties. At MIT 1985 questionnaires disclosed that 11 percent of the faculty[25] and 11 percent of the graduate students[26] had security clearances.

CONSEQUENCES FOR GRADUATE STUDENTS

From the previous discussion, it has already become evident that there are serious consequences for graduate students over which they have little control.

20. James A. Ionson, "SDIO/IST Memo" (Department of Defense, 8 Aug. 1985).

21. *National Policy on the Transfer of Scientific, Technical and Engineering Information*, National Security Decision Directive 189, 21 Sept. 1985.

22. Colin Norman, "Security Problems Plague Scientific Meetings," *Science*, 26 Apr. 1985, p. 471.

23. R. Jeffrey Smith, "White House Issues Secrecy Guideline," *Science*, 11 Oct. 1985, p. 152.

24. Ibid.

25. "Report of the Ad Hoc Committee on the Military Presence at MIT" (Massachusetts Institute of Technology, Apr. 1986), app. A, p. 6.

26. Ibid., app. B, p. 5.

These include changes in the relative availability of support and, consequently, in the research fields open to them and the choices of topic and emphasis within the fields. For students who do not desire to do research related to weapons programs, there is an additional problem. Work in the area of research which interests them most strongly may only be possible with DoD support, forcing a choice between field and conscience. Other possible consequences are the requirement of clearance for U.S. students and the exclusion of foreign students, now a substantial presence in all departments. Outright classification of theses has been rare, but thesis research is being carried out in classified areas. Classification of the actual thesis is avoided by excluding the portions of the research that would make this necessary, resulting in a document that does not adequately describe the student's work.

There are also career implications, since thesis research in DoD-supported areas leads naturally to employment in weapons RDT&E, one of the purposes of DoD support of university R&D. Dr. Robert Rosenzweig, president of the Association of American Universities, gave the following analysis in his testimony to Congress in support of the DoD university research budget:

The point at which career decisions, career directions, begin to be set for graduate students is the point at which they decide what direction they are going to go on their dissertation. If they are engaged early in work that is intellectually stimulating to them and has some promise for the future and is supported by the DoD, it seems to me you are well on the way to having them hooked into that enterprise for a long time.[27]

27. *Department of Defense Appropriations for 1986, Hearings before the Subcommittee on the Department of Defense of the Committee on*

Even if there are other employment options, if the number of students who received their degrees in a particular area is large relative to non-DoD opportunities, then for a substantial number choice may be limited to weapons RDT&E or leaving the field. If the student's thesis is curtailed to avoid classification, it decreases his or her ability to compete for non-DoD-associated positions.

Finally, if national priorities change and the DoD budget decreases, there is the possibility of substantially fewer jobs in weapons RDT&E areas. While it is usually possible to change fields, there is frequently a considerable cost to the individual and a loss of educational investment for the nation.

POLITICAL CONSEQUENCES

Another aspect of the increasing number of degree recipients in areas relevant to weapons RDT&E is that it increases the constituency for the continuation of those programs. There are some scientists and engineers in weapons programs who are willing to take public positions opposing proposals with which they disagree. For example, more than 1600 individuals at industrial, government, and private, not-for-profit laboratories signed a letter to Congress asking that the rate of increase of SDI funding be slowed, since "[a] highly accelerated research program will inevitably result in waste, overburdened management and incomplete technical scrutiny."[28] But others have engaged in active lobbying with Congress for increases in support. The majority are silent, but their numbers

Appropriations, House of Representatives, 99th Congress, First Session (Washington, DC: Government Printing Office, 1985), pt. 8, p. 771.

28. "1,600 Scientists Tell Congress That SDI Carries Funding Risk," *Los Alamos Monitor,* 19 June 1986.

are so substantial[29] that there would be a serious problem with their relocation if the programs in which they work were cut back. This builds in a resistance to funding decreases in these programs.

At the universities, research workers supported by DoD funding are usually silent on the issues relevant to that funding, even if they say that their own research has no connection with weapons RDT&E. One of the reasons given by a member of the MIT physics department for not signing the SDI pledge was that in a few years "everything" would be supported by SDIO/IST and he did not wish to risk retaliation. This caution was not misplaced. In 1986 the under secretary of defense for research and engineering, Donald A. Hicks, made the following statement: "[I]f they want to get out and use their roles as professors to make statements, that's fine, it's a free country, . . . [but] freedom works both ways. They're free to keep their mouths shut . . . [and] I'm also free not to give the money."[30] This bluntness resulted in an outraged response from prominent scientists, including Sidney D. Drell, then president of the American Physical Society, who said: "Will the nation be stronger if these scientists are lost to research or silenced in the public debate of this important issue?"[31] Dr. Hicks subsequently resigned for unrelated reasons,[32] but his point of view has been evident in government before, notably on the part of President Nixon, and certainly continues to be felt.

Acceptance of DoD funding has political consequences, since it is used as evidence for support of the programs, irrespective of the protestations of the recipient. Dr. James A. Ionson, director of SDIO/IST, remarked at the March 1986 meeting with university representatives to introduce that program, "[B]ut this office is trying to sell something to Congress. If we can say that this fellow at MIT will get money to do such and such research, it's something real to sell."[33] Since the universities have not just been passively accepting DoD funding, but have been actively seeking it, it is even more difficult for them to claim disinterest in the R&D programs.

ACADEMICIANS AND
SOCIAL RESPONSIBILITY

Many people are troubled by military funding of university research on ethical grounds, stemming from a judgment that the arms race is not in the interest of national security. In contrast with public service, which is usually considered to be response to a need voiced by the government, social responsibility can take the form of opposition to government policies, as it did in the late 1960s and early 1970s with respect to the war in Vietnam. That period brought protests against military research on campus, and faculty committees were convened to examine the issues. At MIT, the Review Panel on Special Laboratories concluded:

As far as M.I.T. is concerned, the nation's emphasis on defense produces a bias toward specific areas of research at the Institute, and makes it more difficult to move in other directions. M.I.T. has a role to play in attempting to redress this balance, not only within itself but also at the national level.[34]

29. *1982 Postcensal Survey*, tabs. B-12 and B-27.

30. R. Jeffrey Smith, "Hicks Attacks SDI Critics," *Science*, 25 Apr. 1986, p. 443.

31. Sidney D. Drell, "Disloyalty and DoD Funding," *Science*, 6 June 1986, p. 1183.

32. "Comings and Goings," *Science*, 31 Oct. 1986, p. 540.

33. R. Jeffrey Smith, "Star Wars Grants Attract Universities," *Science*, 19 Apr. 1985, p. 302.

34. "The Review Panel on Special Laboratories: Final Report" (Massachusetts Institute of Technology, Oct. 1969), p. 4.

In the 1980s there have again been events that have troubled the universities. The controversy at Cal Tech over the Arroyo Center has already been mentioned, and a particularly thought-provoking case occurred at Stanford University. A weapons-related experiment proposed by Lawrence Livermore National Laboratory for the Stanford Linear Accelerator Center (SLAC) Synchrotron Radiation Facility caused heated discussion of the appropriateness of this proposal and the unwilling involvement of some SLAC personnel in its support. These issues were considered by the Stanford University Faculty Committee on Research, which concluded that end use should not be a university criterion for denying research support and that such considerations were the responsibility of the individual investigator. However, since the protesting SLAC personnel were involved not by choice, but by their functions in the laboratory, attempts should be made to find them other positions at SLAC.

Is it reasonable to say that social responsibility rests solely with the individual investigator and not with the institution? The individual investigator is seldom solitary these days and at universities is likely to have students, postdocs, and technical people supported by his or her research grants. These individuals, like the SLAC personnel, may disagree with the end use of research in which they are involved, but will have only the options of acquiescing or quitting unless the institution takes responsibility for finding them other places internally. Furthermore, the individual investigator is not solely responsible for the funding, and he or she is not the only one to benefit. The universities have been lobbying for DoD support and are therefore implicitly, and sometimes explicitly,

encouraging its acceptance. The universities benefit through overhead payments, institutional grants, and upgrading of facilities and equipment. It is, therefore, not accurate to say that they play no role in the funding of the individual investigators.

However, senior scientists and engineers do have a considerable freedom of choice in where they apply for support and what support they accept. The usual criteria are whether the support will permit research which the investigator wishes to supervise and whether it will advance him or her professionally. The receipt of any type of funding does the latter, although various types vary in prestige. A substantial majority of faculty feel that the origin of the support is unimportant, that money has no color. They would argue that probable end use is not a feasible criterion for much research, that outcomes are not predictable. A good example is the research of Arthur W. Gallstone on the enhanced flowering of soybeans caused by 2,4,5-triidobenzic acid, research which subsequently led to the use of a related chemical as a defoliant in Vietnam.[35] Even though, in the case of research funded by DoD, the purpose is explicit, it is argued that the research is basic and not directly connected with the possible end use. The possibility of negative consequences is dismissed as unimportant or somebody else's responsibility.

It is clear that this pervasive attitude leaves a residual discomfort. This surfaced in the MIT debate over SDIO/IST funding, when those who suggested signing the pledge were accused of violating the academic freedom of their colleagues to carry out research as they saw fit. Phrases like "the tyranny of colleagues" were used to describe the pressure

35. Arthur W. Gallstone, "The Education of an Innocent" (Report, Hastings Center, Sept. 1971).

that was supposedly brought to bear. The reason for the strong reaction was that those interested in SDIO/IST funding did not, in general, support the SDI program but were operating on the Sutton principle. They agreed with the critics of the program but found various rationalizations for personal acceptance of the support. This observation is based on conversations with many MIT faculty members concerning the pledge and on a 1985 questionnaire of the MIT faculty. Only 13 percent of the respondents disagreed with the statement: "[The SDI] is unlikely to result in a useful defense system," and only 5 percent thought that the effect of SDI funding would be generally good for MIT.[36]

The SDI pledge was not a violation of academic freedom; instead it was a breach of collegiality. Academic freedom is the freedom to carry out research and teaching without interference for political or ideological reasons, but it is not freedom from criticism. Rather than mandating collegial silence, it protects the right to criticize.

CONCLUSION

Universities are educational institutions, and, in addition to the formal curriculum, they teach their students how to function as scientists and engineers. The senior research personnel are the models from which the students learn their future roles. The point of view that only success in research and in research funding are important, to the exclusion of other attributes, is often forcefully conveyed, if not always articulated. Excellence in teaching is usually considered a positive, although not necessary, attribute in faculty members, but taking into account the possible end uses of one's

research and the possible consequences of using a particular source of funding is not. This is, of course, not only true with respect to research funded by DoD, but the controversy about the escalation of the arms race and DoD funding of university R&D have focused attention on this in recent years.

What should take place is simple to state but not to effect. Universities should not solicit or encourage funding by mission-oriented sources without a faculty consensus that this is desirable. Individual faculty members should take responsibility for foreseen consequences of their research, including those attached to seeking or accepting support from particular sources. Social responsibility should become important among the criteria of excellence at the universities, a factor in promotion and tenure decisions.

DoD funding of university research has consequences for the university, the students, and the nation, and it has no benefits that could not be achieved with other funding sources. Unless there is a specific and nationally acknowledged urgent reason for the acceptance of DoD support for a particular project, it should not be sought. The universities will produce a new generation of scientists and engineers who can quickly turn to military problems in the event of a national crisis, as was the case in World War II. Both university administrators and research workers should support decreases of DoD funding, at least to a level where it is not a major factor in any field. This should be accompanied by strong pressure on the president and Congress for increased funding through agencies dedicated to the free support of basic research and of applied research relevant to the pressing problems of the civilian economy. Similar attention should be brought to bear to ease various problems experienced by

36. "Report of the Ad Hoc Committee on the Military Presence at MIT," app. A, pp. 7-8.

those in university R&D, such as year-to-year uncertainties and crippling paper-work requirements. Further increases in DoD support have been suggested as a solution to these problems and other congressional vagaries. It is not at all obvious in the current situation that this assessment is correct, but even if it were, the negative consequences of DoD funding are much too great a price to pay.

Book Department

PAGE

INTERNATIONAL RELATIONS AND POLITICS 155
AFRICA, ASIA, AND LATIN AMERICA 160
EUROPE 166
UNITED STATES 174
SOCIOLOGY 183
ECONOMICS 190

INTERNATIONAL RELATIONS AND POLITICS

JONES, ARCHER. *The Art of War in the Western World.* Pp. xix, 740. Urbana: University of Illinois Press, 1987. $34.95.

"In war," wrote Clausewitz, "everything is very simple. The problem is that to do even the simplest thing is very difficult." The same comment might be applied to this impressive volume, which ambitiously sets out to provide an overview of the development of the art and science of Western warfare from the times of the ancient Greeks to the present day. On the one hand, it is relatively simple for the experienced and gifted historian—and Archer Jones certainly qualifies under both descriptions—to describe the salient changes in weaponry, tactics, and operational doctrine that have succeeded one another down the centuries of the last two and a half millennia. On the other hand, it is very much harder in the space of a single volume—even a large one such as this—to analyze the reasons underlying those changes and to appreciate their full implications. Broad generalizations may suffice for many readers, but not for the more specialized seekers after knowledge, understanding, and truth. One great strength of this book is the way it caters to both categories of reader, remaining highly readable and at the same time authoritative.

In the present age of—possibly excessive?—historical specialization, it has become increasingly rare to find a true generalist. Archer Jones is therefore doubly to be commended. His reputation outside America has until now rested largely upon his major contributions to the history of the U.S. Civil War's strategy. To this repute he has now added a powerful demonstration of his skills as a generalist, taking much of mankind's turbulent history as his subject. Clearly, no blinkered one-war historian he.

At one level it is a fairly straightforward history of the development of warfare, drawn mainly, as Jones freely admits, from creditable secondary sources. At another, making the most perhaps of his experience at the United States Army Command and General Staff College, he examines what he terms the changes in "certain operational variables." These variables are linked to military factors, particularly operational aspects, at the expense of political and sociological or institutional approaches, supported by a positively Jominian mass of graphs, diagrams, tables, and schematics that both tantalize and infuriate the reader—as no doubt was their designer's intention.

There are, inevitably, certain criticisms that can be made. Although each chapter is provided with a minimum of annotations at the end of the volume, there is no overall *bibliographie raisonnée,* which a work of this nature really requires. Second, the occasional line illustrations do not add a great deal to the work and their purposes could have been better served, in many instances, by photographs or prints. More important, however, is a certain unevenness in overall historical coverage. Jones seems determined to exercise a strict "self-denying ordinance" where American military history is concerned. The U.S. Civil War, for example, is not treated at all, and the American Revolution earns only the barest mention—just two entries in the index and those divided by 100 pages. Jones is entitled to select his areas, but it is again surprising to find no mention of General MacArthur or the Pacific War. Admittedly, the struggle against Japan did not take place in the Western hemisphere geographically, but Western military influences were surely paramount on both sides. Similarly, neither Korea nor Vietnam appears. On the other hand, Jones has much time and space for the Campaign of 1940 in France and the Low Countries. These peculiarities of omission and commission must lead one to challenge the full validity of the dust jacket's claim that this volume "is likely to stand for a generation as the standard work on how men conduct war." It should, perhaps, read, "how some men conduct some wars."

In the final analysis, what we have here is an expert consideration and analysis *in extenso* of the European art of war from the earliest times, rather narrowly defined though that area proves to be. Jones may have deliberately wished to make a clear distinction between the military experience of the Old World and that of the New, but, if so, he could have included a fuller justification. It is true that in many ways European experience and doctrine were the twin crucibles of the Western world's art of warfare—until at least the onset of nuclear weaponry—but Archer Jones's Eurocentric preoccupation is possibly excessive.

Nevertheless, there are many excellent things in this *jeu d'esprit,* and even if not every reader will agree with Jones's total vision of matters military, one thing is absolutely certain: this book will make the reader think.

DAVID G. CHANDLER
Royal Military Academy Sandhurst
Camberly
Surrey
United Kingdom

MODELSKI, GEORGE and WILLIAM R. THOMPSON. *Seapower in Global Politics, 1494-1993.* Pp. xii, 380. Seattle: University of Washington Press, 1988. $35.00.

During the last decade, George Modelski and William R. Thompson have assembled a great deal of evidence for the existence of a long cycle of world politics, which they define as "the process of fluctuations in the concentration of global reach capabilities which provide one foundation for world leadership." These long cycles, which have recurred with some degree of regularity, have been punctuated by global wars—during the years 1494-1516, 1580-1608, 1688-1713, 1792-1815, and 1914-45—that have laid the foundation for successive phases of Portuguese, Dutch, British, and American global leadership.

The present book has two objectives: first, to propose a conceptual model of power in the modern world system based on command of the sea; second, to develop, through exhaustive research on navies, an empirical base enabling Modelski and Thompson to devise a scale of global power at any given time.

Following Mahan, Modelski and Thompson contend that sea power constitutes the central element of global power. Unlike land armies, navies confer greater mobility, which is essential to global leadership, and are more closely related to general technical innovations. Part 1 of the book, an extended review of naval history, provides evidence for two hypotheses: first, that world powers have been sea powers; second, that fundamental changes in the balance of power are associated with changes in the distribution of sea power.

The evidence adduced by Modelski and Thompson is persuasive. Global wars have been naval wars, and decisive battles have been waged for control of the sea lanes. Moreover, questions of sea power have been important immediate causes of war. This is, perhaps, more obvious for the years prior to the twentieth century, but even in this century, continental powers, including Germany, have proved unable, in the absence of naval superiority, to translate their economic and military predominance into global leadership, and the same might be said of the Soviet Union.

Having demonstrated the centrality of sea power, Modelski and Thompson proceed to operationalize naval supremacy in terms of possession by one state of 50 percent or more of the capital ships. Part 2 includes extensive documentation, including a time series of global concentration, of the navies of the global and middle powers. Using the 50 percent benchmark, Modelski and Thompson find a close correspondence between their hypothesized long cycles and naval supremacy in the cases of Portugal (1494-1540), the Netherlands (1594-1642), Britain (1714-50, 1836-60), and the United States (for the years projected to be 1946-93).

The empirical analysis presented in this book represents a significant contribution to the analysis of long cycles and will also be useful to scholars of maritime history. In operationalizing and quantifying global power in terms of naval superiority, Modelski and Thompson seek to give precision to a concept that has often been treated rather loosely in the study of international relations. In defining power so narrowly, however, they also run the risk of ignoring political and economic factors that not only constitute essential elements of a nation's power base but also might be included in a more comprehensive explanation for the existence of long cycles.

ALAN W. CAFRUNY

Hamilton College
Clinton
New York

REISS, MITCHELL. *Without the Bomb: The Politics of Nuclear Nonproliferation.*

Pp. xxii, 337. New York: Columbia University Press, 1988. $35.00.

Since the advent of the nuclear age, controlling weapons of mass destruction has been the overriding concern of strategic policy. Attention has focused mainly on the risks of war between the superpowers and on the means for preventing it. By contrast, concern about the spread of atomic arms waxes and wanes. The slow pace of proliferation, and confidence in the effectiveness of a nonproliferation regime built around the Nonproliferation Treaty and the safeguard system of the International Atomic Energy Agency (IAEA), have served to assuage anxieties. Only the threat of a new entrant to the nuclear club focuses public interest.

Yet the proliferation danger remains real, and there is a good chance that it will quicken. It is opportune, therefore, to have so excellent a reappraisal of proliferation issues and prospects as that presented by Mitchell Reiss. A thorough and knowledgeable assessment of six national nuclear programs—those of India, Israel, Japan, South Africa, South Korea, and Sweden—constitutes the heart of the book. But it is much more than a compendium of policies and plans, for Reiss has fixed on the crucial issue of proliferation, namely, what the incentives and pressures are that motivate a state to consider taking up the nuclear option, and on its companion question, How influential are the restraints and disincentives created by IAEA safeguards?

The evidence that Reiss adduces leads him to some noteworthy conclusions. Above all, as is indicated by the subtitle of his book, he stresses the political factor. Security worries impel the interest in nuclear arms. Therefore, security alliances and commitments should be seen as an integral part of the nonproliferation regime. The restraints recognized and observed by countries as diverse as South Korea, Sweden, and India derive primarily from their security environments, including, *inter alia,* nuclear guarantees from protective allies, the risk of provoking countervailing actions, opportunities to meet military needs by conventional means. The bilateral ties with a nuclear-competent protector are of especial importance. Indeed, as Reiss perceptively notes, "the U.S. link unwittingly encouraged a few countries to threaten to acquire nuclear weapons" since that threat itself generates reaffirmation of American security commitments.

IAEA safeguards, for their part, are barriers, not effective restraints. Together with the Nonproliferation Treaty they should be viewed as "earnests of intent" that "reinforce other sources of restraints"—mainly security ones. Therein lie both risk and opportunity, for that circumstance highlights how essential it is for the United States to maintain its overseas commitments, and the implications of failing to do so. It also poses a formidable challenge to American diplomacy to "calibrate specific policies to the particular circumstances and concerns of each country" based on an understanding of the "idiosyncrasies of potential nuclear weapons states."

MICHAEL BRENNER
University of Pittsburgh
Pennsylvania

VIGOR, P. H. *The Soviet View of Disarmament*. Pp. vii, 189. New York: St. Martin's Press, 1986. $25.00.

BLACKER, COIT D. *Reluctant Warriors: The United States, the Soviet Union, and Arms Control*. Pp. xiii, 193. San Francisco: Freeman, 1987. Paperbound, $12.95.

Vigor's book deals with proposals for disarmament and arms control as proposed by the Soviet government to other countries or to international organizations and meetings. Vigor has ensured that the views discussed are official and authoritative by including only the actual proposals or the views of the Politburo.

The book goes into somewhat detailed descriptions of the proposals from 1917 to 1980. The main proposition of the book is that the Soviets, following Lenin's thesis, believe that militarism is an inherent characteristic of capitalism and that no useful agreement can be achieved by them with the capitalist coun-

tries. Nevertheless, it is considered to be in the interest of the Soviet Union to retain proposals for disarmament in case, by some miracle, the capitalists agree to sign them. Vigor sees certain patterns in the Soviet proposals. Before World War II, the Soviets were more interested in total disarmament. After the war, however, they favored more partial-disarmament proposals. They have not presented any total-disarmament treaty since 1965.

Vigor discusses almost all the actual agreements: the Partial Test Ban Treaty and both Strategic Arms Limitation Treaties. The problem here is that the materials are now eight years old. The Intermediate-range Nuclear Forces (INF) Treaty, the negotiations for which, Vigor says, are stalled, has been signed. In several instances Vigor says that the Soviets would never allow on-site verifications; the pictures of smiling Americans arriving in Moscow to inspect the Soviets dismantling their missiles, as agreed upon in the INF Treaty, must be embarrassing to him. The work might still be a valuable source for the historians interested in the subject. It has ample references and historical anecdotes.

My problem with the book is that it reads like a satire on the international relations authors of the realist school such as E. H. Carr or Hans Morgenthau. Examples include sentences such as "I may appear to portray the Russians as being motivated in the field of disarmament by consideration of the Soviet self-interest above all else" or "The primacy of Soviet security as a determinant of Soviet disarmament policy has been clearly demonstrated several times during the course of this book." But what types of proposals should the Soviets have advanced—those that would benefit the capitalist states or be detrimental to Soviet interests?

Blacker's book concerns the impact of nuclear weapons on the relations between the Soviet Union and the United States. He argues that nuclear weapons have radically changed the concepts of offense, defense, and deterrence. The book is a well-written history of the relations between the two superpowers since the dawn of nuclear weapons. As such, it is an excellent textbook for undergraduate courses on the arms race, foreign policy, or international relations. It has a chronology, and a glossary to explain the jargon of superpower relations.

The book seems to be overly pessimistic because it does not see the possibility that the Soviets might accept Reagan's zero option in the INF negotiations. It seems the tricky Soviets pleasantly surprised both Blacker and Vigor. They agreed to what is essentially Reagan's proposal with the inclusion of the missiles in Germany. Blacker advises the Americans to expect arms reductions but not disarmament for the foreseeable future. He also believes the solutions to the threat of nuclear war are political, not technological wizardry like the Strategic Defense Initiative.

A. REZA VAHABZADEH

University of Pennsylvania
Philadelphia

WOOLEY, WESLEY T. *Alternatives to Anarchy: American Supranationalism since World War II.* Pp. xi, 244. Bloomington: Indiana University Press, 1988. $25.00.

Alternatives to Anarchy is a timely study of three recent movements in the United States that sought to transcend the limitations and constraints of nationalism. The commanding rationale behind each was the conviction that the nation-state could no longer cope with the dangers and problems of the modern world. These attempts at supranationalism originated out of specific fears that seemed to many to threaten the security of the United States as well as the continuance of life on the planet.

First was a world government movement that promoted the immediate creation of supranational government. This call for world government was motivated largely by the worldwide insecurity accompanying the advent of the atomic era. The movement peaked between 1947 and 1949 and declined rapidly thereafter, largely a victim of the developing cold war, a low level of public support, vagueness in outlining the guidelines of world government, and the inability of Americans to transcend national self-interest.

Another attempt to move beyond nationalism sought union between the Atlantic democracies. It emerged frankly as a response to the cold war and the belief that world government was not currently achievable, although many hoped it might constitute a first step in that direction. The movement was spearheaded by the Atlantic Union Committee, an effectively organized citizens' committee that made a strong bid to gain support but "faltered just as it appeared to be on the brink of success." The Atlantic Union Committee remained "narrowly elitist" and thus failed to develop support among the voting public.

A third attempt to push beyond national boundaries Wooley calls "functional supranationalism." This name refers to united efforts to deal with world problems through cooperative agencies and arrangements across national boundaries. Such problems as arms limitation, protection of the environment, population control, and standards for international trade might be handled in this fashion. Again proponents hoped this movement would evolve toward greater political unity. But even though many supranational projects and agencies continue to operate and probably always will, the larger objectives of the movement, an expanding spirit of cooperative endeavor, were not achieved.

Wooley's excellent study traces the failures of supranationalism in modern times. He writes, however, from a position of sympathy for the movement, now confident that only some great "traumatic experience" can provide the thrust leading to greater supranational unity. Even so, the book reveals little of detectable bias. It is thoroughly researched in both secondary and primary sources, and the data are subjected to convincing analysis. It is a delight to read, despite the somewhat technical nature of the subject. The book is a must for students of the cold-war era.

FRANK D. CUNNINGHAM

Southwest Baptist University
Bolivar
Missouri

AFRICA, ASIA, AND LATIN AMERICA

BANISTER, JUDITH. *China's Changing Population*. Pp. xv, 488. Stanford, CA: Stanford University Press, 1987. $49.50.

This substantial opus, based on a 1977 doctoral dissertation from Stanford University, sets a new standard against which modern Chinese demographic studies will be measured. It deals with the period from roughly 1949, when the People's Republic of China was founded, to 1985, and it has been made possible in its present form by the transformation in the quality of population data on China since 1982, when the People's Republic conducted the third of its censuses, one that is reputedly far more comprehensive than the earlier two in 1953 and 1964, respectively.

The work is obviously of interest not just to demographers and China scholars but also to development economists and sociologists, who must accept Banister's contention that "if we can ascertain China's demographic trends, we have added the largest missing piece to our knowledge of world population dynamics." From my admittedly nontechnical perspective, one of the central questions is that of comparing India's failure to China's success in the control of fertility since 1950 and more conspicuously since about 1970. These questions are addressed in a brief introductory chapter and in the conclusion, although evidently chapter 7, on the so-called one-child family campaign, is also of some interest in this context.

After briefly outlining the history of China's population from roughly A.D. 2, the date of the first census of the Han dynasty, to 1953, Banister sets out to discuss the quality of the population data collected since 1949. The historical discussion is based largely on John Durand's 1960 paper in *Population Studies*. One puzzling omission from this section, as indeed from the book's quite extensive bibliography, is Ping-ti Ho's *Studies on the Population of China, 1368-1953* (Harvard University Press, 1959).

The core of the book—chapters 2 to 9— deals with health and morbidity, mortality and fertility, nuptiality and family planning, as well as spatial issues such as density and migration. A particularly interesting discussion is that in chapter 9, on ethnic groups. Here Banister demonstrates that while the relatively small non-Han ethnic populations continue to be resident in their "historical homelands in the mountainous and the arid parts of the country," various post-1949 processes, including the promotion of internal migration and the redrawing of provincial boundaries, have tended to increase Han control over these very areas.

Of particular utility is Banister's extensive and impartial discussion in chapter 7 of the one-child family campaign, this campaign being regarded by many as central to the ongoing process of reducing the population growth rate in China. Banister concludes, quite convincingly, that the program does contain significant doses of coercion and is also likely to exacerbate rural-urban differentials, given the differential structure of incentives. Most disturbing of all is the suggestion that the policy may lead to the creation of "two classes of children born in the 1970s and thereafter," a privileged class of single children and a lower class of children with siblings. Other implications of this policy include potential increases in female infanticide, due to strong male-child preferences still existent in rural Chinese society. The discussion of eugenics as practiced in China is also likely to be an eye-opener for some of the People's Republic's more unabashed admirers.

It is clear that Banister herself is considerably troubled by the various ethical questions involved, as witness her statement that "journalists and scholars who research, write about, and speak about the demography of China have a moral and ethical obligation not to gloss over the compulsory and coercive qualities in China's family planning program, so that their audiences can reach an informed judgement." No matter what one's stance on such questions may be, however, one cannot gainsay the book's considerable technical achievement in collating data at a local, provincial, and national level to provide the reader a state-of-the-art treatment. Some eighty tables and several figures should satisfy even the most ardent data cruncher. This book may well remain the standard reference for several years to come on this topic. It has the advantage of being accessible, in terms of substance and argumentation, to the intelligent undergraduate, at the same time preserving a relationship to primary material that the puristic researcher will find hard to fault.

SANJAY SUBRAHMANYAM
University of Pennsylvania
Philadelphia

BULMER-THOMAS, VICTOR. *The Political Economy of Central America since 1920.* Pp. xxiv, 416. New York: Cambridge University Press, 1987. $49.50. Paperbound, $16.95.

In this book, Bulmer-Thomas uses his previously unpublished estimates of the national accounts to explore economic and social development in the five Central American countries from 1920 to 1987, and to shed light on the current crisis. Even though many of the judgments in the book are backed with quantitative material, the text is pleasant to read and the tables and graphs easy to understand. It is somewhat like a history book based on clear, hard data and facts.

The words "political economy" in the title allow Bulmer-Thomas to link economic development and political change in the region. Economic events, such as industrial development, agricultural output, regional organizations, and so forth, are presented in connection with the prevailing political events. Bulmer-Thomas feels external influence on Central American events has been greatly exaggerated. Therefore, less emphasis is put on external intervention and foreign aid, and more on internal affairs.

The book starts out in 1920 with the development of the export sector in Central America. Bulmer-Thomas believes that the success of economic and social reform in Central America has been heavily influenced by a dominant export-led model. This model, described in its evolving phases, is perhaps overstressed throughout the book.

The 1929 depression provoked a fiscal crisis and contributed to a significant increase in social unrest with a decline in real incomes. Government attempts to reduce expenditure to match the fall in revenue were not successful, and budget deficits were common in the first years of the depression. With political repression on a massive scale, stagnation in the traditional export sector, and an unwillingness on the part of authorities to countenance an alternative to the export-led model, the decade of the 1930s earned a reputation as the dark age of Central American history. By 1944, the democratic reform movement had shaken the foundations of the authoritarian *caudillismo* to its roots. The war years brought Central America and the United States closer together in strategic and economic terms, and U.S. direct foreign investment rose significantly.

Central America came out of World War II with an economy that exhibited many of the classic signs of underdevelopment. Exports continued to be dominated by earnings from coffee and bananas. At the end of the war, the economies of Central America faced the task of capitalist modernization. This involved creating an environment in which capital accumulation could take place outside the traditional export sector and breaking down obstacles created by a weak financial system, a highly unequal pattern of income distribution, and a poorly integrated internal market.

By 1954, the Central American economies had undergone a series of transformations that left them in a much stronger position to withstand adverse external conditions and that would have a profound impact on the social and political developments of the region for the next 25 years. El Salvador, Guatemala, and Nicaragua were condemned to reactionary despotism, while Costa Rica and Honduras offered their citizens the prospect of a more democratic world.

The formation of a regional common market in the 1960s, coupled with exceptionally generous fiscal incentives for industry, created interest among Central American investors. By 1970 some 70 percent of industrial production was not controlled by foreigners, with the food sector accounting for 50 percent of gross industrial output. Central America was hit by three serious external shocks: inflation, the oil crisis, and a set of natural disasters.

Particular emphasis is given to 1979, which, with the fall of Somoza in Nicaragua, the collapse of General Romero's regime in El Salvador, and the beginning of the worst economic crisis since at least 1929, is a year of very special significance. Bulmer-Thomas wraps up the book with a conclusion that brings together the analysis of the previous chapters to summarize the main findings and provide certain conclusions, followed by a statistical appendix.

All throughout, the reader can observe that the region exhibits both conformity and diversity—economic unity due to common external influences, filtered through domestic institutions to affect each economy in slightly different ways. Bulmer-Thomas does a fine job in presenting these economic, social, and political similarities and differences in a comprehensive, yet easy-to-follow, history of the Central American region since 1920.

JOHN C. BEYER

Robert R. Nathan Associates
Washington, D.C.

EVANS, PAUL M. *John Fairbank and the American Understanding of Modern China.* Pp. xvi, 366. New York: Basil Blackwell, 1988. $24.95.

Evans's biography of China scholar John Fairbank provides an excellent companion piece to Fairbank's own autobiography, entitled *Chinabound: A Fifty Year Memoir,* which was published in 1982 by Harper & Row. Evans began researching him in 1977, just as Fairbank reached his seventieth birth-

day and retirement from his teaching career at Harvard, which freed him to do the autobiography. That the autobiography was published six years before Evans's biography is an indication of the careful attention to documentation and accuracy in the latter. Indeed, as sometime Fairbank rival Lucian Pye puts it, "Evans' biography of Fairbank tells more about his life and his thinking than does Fairbank's own autobiography" (see the back cover of the Evans volume).

This is not to say that Evans reveals any deep dark secrets that the autobiography seeks to cover up. The biography is, in fact, more complimentary to its subject than the autobiography, wherein Fairbank confesses somewhat obliquely to being more of an academic "organizer" than a scholar. From Evans's study he emerges as a hero, at least insofar as an "organizer" can be a hero. Evans began his study as a doctoral dissertation. His original plan was to study several American interpreters of modern China, such as Owen Lattimore, Edgar Snow, George Taylor, Fairbank, and others, but he clearly became so fascinated with Fairbank that he has relished spending a decade studying his career.

A keen and assiduous documentarian, Evans utilized some 260,000 items contained in Harvard's collection and Fairbank's own private collection, and he seems to have tracked down almost everything Fairbank has ever written or said publicly. The only sources he had problems with, and he admits this, were some of the Federal Bureau of Investigation's files on certain security problems Fairbank had—passport denial and the like—during the 1951-52 McCarthyist attacks on the Institute of Pacific Relations, with which Fairbank had minor connections.

The "slings and arrows" (chapter 6) that Fairbank incurred as a result of these and of his own disillusionment with both the Nationalists and their Kuomintang, which began with his experiences in wartime Chungking in 1943 (chapter 4, "At War"), and the Communists after Mao's triumph in 1949 (chapter 5, "The Social Revolution") led him away from political involvement and into what Evans calls his program of "National Enterprise National Interest" (chapter 8) for China studies. This meant coordinating all possible elements of academia, government, and private foundations in a grand China-studies effort. This activity required considerable academic politicking, such as convincing classical Sinologists that modern China was worth studying, China, or Kuomintang, lobby advocates that he was not pro-Communist, and anti- Vietnam-war activists, who included some of his own students, that he was not allowing the Central Intelligence Agency into his Harvard program—which he was. His approach to China studies was, Evans explains, non- and even anti-ideological. Thus, for pragmatic reasons, he sought and accepted funding from both private and government sources, and good students from whence they came, including State Department and Central Intelligence Agency personnel.

He could, of course, be criticized for this, and indeed, my senior colleague at the University of Pennsylvania, Professor Derk Bodde, led us to refuse to become a National Defense Education Act "center." I joined Bodde in this, for I still think such refusniking helped to end McCarthyism and the Vietnam war. But, on the other hand, Harvard's East Asia Library and programs have dwarfed Penn's by comparison. Evans is very even-handed in his treatment of such matters, and he concludes that what may well be called Fairbank's crusade for China studies has magnificently paid off in developing the "American connection" (chapter 12) with the new China in recent years and should be adjudged well worth the ideological compromises made along the way.

HILARY CONROY
University of Pennsylvania
Philadelphia

LEWIS, BERNARD. *The Political Language of Islam*. Pp. vii, 168. Chicago: University of Chicago Press, 1988. $14.95.

MUNSON, HENRY, Jr. *Islam and Revolution in the Middle East*. Pp. xi, 180. New Haven, CT: Yale University Press, 1988. $18.95.

The Political Language of Islam is the revised version of the Exxon Foundation Lectures that Bernard Lewis delivered at the University of Chicago in 1986. In the course of five chapters, Lewis discusses aspects of Islamic political language: "Metaphor and Allusion," "The Body Politic," "The Rulers and the Ruled," "War and Peace," and "The Limits of Obedience."

Attention to the political language of Islam is a relatively recent phenomenon. Although Islamic political and social theory have long been studied by Rosenthal, Lambton, Launst, and others, the specific attention to the nature of political language in Islam is limited to such studies as those of Michael Fischer, Bill Hanaway, and Peter Chelckowski, who by concentrating on the various symbolic uses of words and images in the Iranian revolution have discussed certain related issues in their modern contexts.

Lewis's point of departure is modern uses of the Islamic political language. Here, although Lewis brings much of his historical and linguistic erudition to teach us the inner workings of a religiously charged political discourse, he, alas, shares with other distant students of the Iranian revolution the almost total neglect of the full range of political languages that collectively contributed to the organization and outcome of the movement.

Other than the Islamic discourse, liberal democratic and radical revolutionary discourses, both Western in origin, have been crucial in the making of much that has happened in the Muslim world. But the very designation of "the Muslim world" tends to blind us to the variety and complexity of political languages in these societies. To be ignorant of the presence and intensity of these patently non-Islamic—and nonreligious, for that matter—modes of discourse not only distorts the full image of how political messages are communicated, but because of their catalytic impact on the Islamic political language itself, it creates a methodological barrier between the realities of the Muslim world and their students.

As to Islam itself, Lewis again begins with the common but mistaken notion that in "classical Islam there was no distinction between church and state." The historical fact is, however, that church and state, if we take them in their general and not Christian-specific meanings—that is, institutional expressions of religious and political authorities—have always constituted two separate social organizations. Despite their common root in the Koranic revelatory language, the two institutions of *ulama* and *khulafa* have constituted, historically, two distinct centers of authority, subject to a variety of interactions. (For a detailed discussion of this point, see my "Symbiosis of Religious and Political Authorities in Islam," in *Church-State Relations,* ed. Thomas Robbins and Roland Robertson [New Brunswick, NJ: Transaction Books, 1987], pp. 183-203.)

Moreover, Lewis totally identifies what he terms "we in the Western World, nurtured in the Western tradition," with Christianity and Christendom. This, of course, is to subtract half of the Bible, namely, the Old Testament, and to eliminate the significant share of Judaism in the making of "we in the Western World, nurtured in the Western tradition." Neither in the Judaic tradition nor even in the Crusade and Counter-Reformation episodes of Christianity is the separation of church and state in the West as clear-cut as Lewis would like us to believe.

Lewis makes extensive references to metaphors and allusions in various Islamic languages. These are very general yet suggestive references. Although much of the discussion about metaphors of space, time, birth, rebirth, and the like is general and lacks specific textual references, it does point out an innovative and potentially insightful approach to the Islamic political language. Because Lewis remains extremely vague and general, it is very easy to refute much of what he argues. For example, he maintains that "movement upward or forward indicates improvements, while movement downward or backward indicates deterioration or loss of power, status, etc." This cannot always be true. According to

a prophetic tradition, every successive generation after that of the Prophet shall deteriorate in moral dexterity. This would be a corruption in forward movement. The whole doctrinal belief of Muslims in *maad* ("resurrection") is a belief in a return to the point of origin, which is a backward movement. Lewis also maintains that "the imagery of birth and rebirth, renaissance and resurrection, so important in Western usage, has no place in classical Islamic times," whereas there are hundreds of lines of poetry in Arabic and Persian expanding the two images of birth and rebirth stated initially in two famous prophetic traditions: to die before one's death, and to be born twice.

Lewis draws attention to two much neglected areas of political discourse in Islamic literature, namely, *adab* and *fiqh.* Although his assessment of Islamic peripatetic philosophy, that it "flourished for a time in the Islamic academies during the Middle Ages, but it died out and had only a limited impact on later generations," is wrong and outdated by at least half a century of scholarship—even in European languages—on post-Averoës peripatetic Islamic philosophy, Lewis's discussion of some key terms, such as *umma, wilaya,* and *dawla,* is quite informative for lay readers. Equally informative is the discussion of the Muslims' sovereign titles in chapter 3. This chapter is particularly important because of its panoramic view of titles of authority. The chapter on war and peace expounds the crucial theme that there is no precise Arabic equivalent for "holy war." Here, however, Lewis makes the incongruent comparison between Islam and the West, one, in his judgment, having only sacred law, the other both sacred and secular, as if a geographical designation—the West—and a faith—Islam—are logical counterparts.

Although Lewis is right in his final assessment that the Khomeini revolution "owes an unacknowledged debt to the Westernizers and secularists of the past century," he himself, as testified through this book, remains totally uninformed of the nature of this debt. There is scarcely a reference in this book, which begins and ends with a reference to the Iranian revolution, to any seminal thinker whose ideas were instrumental—nay, constitutive—in the making of the modern Islamic political language.

Munson, too, begins and concludes with the Iranian revolution. This book is divided into three parts. Part 1 discusses the subject conveyed by its title, "Islam as Religion and Ideology." Part 2 examines four case studies, of Iran, Saudi Arabia, Egypt, and Syria. Part 3 provides certain explanations of revolutionary events in terms of modernization and resentment of foreign domination.

Part 2, with the exception of chapter 10, is a simplistic regurgitation of the main historical events in the nineteenth century. The four chapters in this section of the book add nothing to our present knowledge about Iran, Saudi Arabia, Egypt, or Syria.

Chapter 9, "Social Bases," is a general classification of social strata into students, the new middle class, the *ulama,* and so forth. Since these very general and categorical classifications are entirely based on secondary sources, this chapter, too, offers no empirical or analytical addition to what we already know.

Thus all of part 2 is a total waste, with the probable exception of its use for undergraduate introductory courses. But since there are many such introductory texts on the market, this, too, will much depend on Yale University Press and its marketing department.

Equally useful for introductory undergraduate courses is the bulk of part 1, a watered-down version of the Sunni-Shiite distinction in Islam, this time cast in the "Islamic sacred history" jargon. This part, too, is based entirely on secondary sources.

What Munson finally has to teach us rests on the last chapter of the book, the third chapter of the third part, "Conclusion: Why Only in Iran?" This ex post facto account of the revolution, through a joggling of "economic crisis," Carter's human rights policy, and Khomeini's charisma, would have probably made a good ten-page article for the *Middle East Journal.* Its projection into a full-fledged 180-page book is a symptom of an inflated market that breeds on mediocrity and more and more resembles journalism, hard-covered into scholarship.

HAMID DABASHI
Harvard University
Cambridge
Massachusetts

STIVERS, WILLIAM. *America's Confrontation with Revolutionary Change in the Middle East, 1948-1953.* Pp. 132. New York: St. Martin's Press, 1987. $24.95.

KIRISCI, KEMAL. *The PLO and World Politics: A Study of the Mobilization of Support for the Palestinian Cause.* Pp. 198. New York: St. Martin's Press, 1987. $32.50.

Kemal Kirisci's study of the Palestine Liberation Organization (PLO) outlines the process whereby that organization gained acceptance in the world arena. William Stivers's overview of American policy in the Middle East demonstrates our reluctance to accept indigenous nationalisms, including Palestinian, as forces to be reckoned with. They share a common theme, U.S. determination to exclude the PLO from any Arab-Israeli peace settlement.

Kirisci's book, the outgrowth of a doctoral dissertation, examines how the PLO, with Arab state support on occasion, changed international consideration of the Palestinian question from that of a refugee problem to one of self-determination. He considers the issue according to groups defined as Arab governments, Third World countries, Eastern and Western Europe, and the United Nations; the United States is included in the chapter dealing with Eastern and Western Europe. In each case, Kirisci treats the question historically and in light of mobilization models to trace changes in regional attitudes.

Kirisci's primary concern is how and when other parties came to acknowledge the PLO, the conducive-environment question, not how the PLO actually mobilized support. He provides a detailed overview of the process and notes that his conclusions are subject to reservations: his use of mobilization models required that "all actors, governments and the PLO alike, [be] treated as though they were unified." This naturally omits consideration of decision-making processes within the PLO that would take note of factional disputes. Kirisci has written an adequate survey of the problem for the nonspecialist, analyzing regional and international factors with care but relying solely on secondary sources. His tables and appendixes will be useful to those interested in the timing and scope of recognition of the PLO and to specialists in mobilization theory.

Stivers's sweeping survey of 35 years of American Middle East policy in 106 pages of text has little room for theory. He contends that American analysts, civilian and military, have always recognized the force of indigenous nationalisms in the Middle East but have consistently opted to put the United States in anti-Soviet alignments with conservative states to preserve Western military dominance in the region. Acting out of concern for short-term expediency, the United States has refused to come to grips with regional problems, especially the Arab-Israeli confrontation, except within the global context of American-Soviet rivalry for control of the area.

I agree with Stivers's basic thesis but believe it conflicts with another proposition that he offers, namely, that U.S. military and political thinking has usually been dominated by an Indian Ocean focus in which the Persian Gulf and its strategic implications, again in an anti-Soviet framework, have often outweighed the Arab-Israeli clash in the minds of American planners. He uses Navy Department and National Security Council documents to buttress that contention.

Stivers has written a trenchant indictment of American foreign policy in the Middle East that reflects both the virtues and drawbacks of his sources. He demonstrates an American interest in the Indian Ocean as a Western sphere that goes back to the 1950s and led to U.S. absorption of Diego Garcia. On the other hand, his argument that the Indian Ocean-Persian Gulf axis has often led to a downplaying of concern for Arab-Israeli matters is dubious, however true it is for the early years of the Reagan administration. He generalizes the case, which relies too heavily and selectively on Navy Department analyses, with no indication of what impact they had on policy formulation. Also doubtful is his statement that the Carter administration floated the idea of a security framework after Camp David that would include Israel, Egypt, and conservative Arab states in a manner similar to the strategic-consensus formula of Haig and Reagan. His reference for this assertion does not support it.

In sum, Kirisci relies totally on secondary materials while Stivers overdoes the significance of those primary documents he employs. Neither book is entirely satisfactory, but both can be read with profit, subject to the reservations noted.

CHARLES D. SMITH

San Diego State University
California

EUROPE

ASCHER, ABRAHAM. *The Revolution of 1905: Russia in Disarray.* Pp. xii, 412. Stanford, CA: Stanford University Press, 1988. $39.50.

It is ironic that the revolution of 1905, one of the most important events in modern Russian history, presents nearly insurmountable difficulties to historians and scholars. It was a troublingly amorphous affair. Anyone attempting to chronicle its course finds it almost impossible to reduce to order, with its spontaneous strikes, widespread agricultural outbreaks, scattered violence among non-Russian nationalities, occasional military mutinies, and sometimes student protests, accompanied by frequent petitions to the czar, to which Nicholas responded with simultaneous concessions and repressions. This "unplanned, unorganized, and unpredictable" series of events produced few memorable dates and heroes. For many months, it did not seem like a revolution at all.

In putting it all together, Abraham Ascher has written a splendid book. This is a volume about what happened in Russia in late 1904 and 1905, and because of Ascher's dedication to detail, it becomes an exciting historical account. Ascher covers all kinds of events, in major cities and remote provinces, enlarging on episodes we have heard of and acquainting us with some we have not. The text is lively with people: with men obtuse and perceptive, well intentioned and arrogant, demagogic and hesitant, with the fearless and the nervous who fumble the ball. We are offered no simplifications, summaries, or categorical answers. Instead, Ascher emphasizes the spontaneity, muddiness, and ambiguity of events; he gives us history, as he says, as it really happened rather than history that well-organized historians might wish to find. As in real life, people are confused, and few are successful. If the concluding chapter analyzes the government's inability to grasp what was happening, it points up the opposition's failures as well.

To a scholar in the exact field, this work may not present data startlingly new. Ascher has utilized published sources, of which there are a multitude—documents, memoirs, histories—supplementing them with newspaper comments and descriptions and with diplomats' notes. Still this is a book of first-rate scholarship, proof positive that one need not always burrow in secret archives to accomplish highly significant historical work.

Ascher believes the revolution continued on for a year or more after the traditional 1905. We are indeed the winners in his analysis, for we are promised another volume, dealing with 1906-7, that is to come.

DEBORAH HARDY
University of Wyoming
Laramie

COHEN, MITCHELL. *Zion and State: Nation, Class and the Shaping of Modern Israel.*

Pp. 322. New York: Basil Blackwell, 1987. $24.95.

There is no dearth of books on the origins of Zionism or the stages of adjustment that enabled the state of Israel to claim a high degree of continuity between ideological definitions of Jewish nationalism and the product created in 1948. Although this book provides yet another general introduction to the varied climate of ideas that preceded and followed the founding of the World Zionist Organization in 1897, its focus on a closely defined ideological controversy—between socialist labor and rightist revisionist views—makes the overall treatment impressively complete in historical detail. At the same time, its hypotheses and conclusions are extremely relevant to contemporary politics in Israel.

Cohen's exploration of the basic principles of the early Poale Zion movement reveals very clearly that this workers' current in Zionist political strategy would place maximum emphasis on the use of national or public capital to build the essential new institutions that the future Israel would need to realize the ideal goals of a true national polity. Because the ideological sources of Poale Zion—and the eventual Mapai Party—were socialistic in origin, such from-the-ground-up institutions—exemplified in the most complete form by Histadruth—reflected a wariness of corruption of the Zionist polity by class interests or influences that could come from imported sources. Those familiar with the delicate balance governing Chaim Weizmann's management of the Jewish Agency just before and after 1930—with its apparent need to convince Diaspora Jews to support Zionist endeavors materially—will realize how socialist ideology had to be adapted to allow the dominance of Mapai political philosophy to become established in the 1930s. This was done through what Cohen calls "segmented pluralism."

For Cohen, these adaptations, by abandoning the necessary principle of the undisputed "identity of the interests of the working class and the nation," were the first steps toward the "reification of Zionism." This process, which would make room ideologically for acceptance of class, state, and nation as separate, if not necessarily opposed, categories, became an essential part of the Labor Party's efforts to combat the political influence of revisionism in Zionism.

Cohen's discussion of revisionist leader Vladimir Jabotinsky concentrates less on the notorious extremist legacy of the Zionist far Right, which spawned the Irgun terrorist organization, than on its ideological challenge to the Mapai Party of Ben-Gurion. Essential to the Right-Left debate were obvious revisionist preferences for a *Mittelstand* rooted in European bourgeois values that would favor private commercial capital over agriculturally oriented socialist labor programs. As important was Jabotinsky's insistence that Zionist objectives were more likely to be attained by international political posturing—something akin to Herzl's original version of "political Zionism"—than by labor's dedication to the gradual construction of an ideal polity "settlement by settlement, immigrant by immigrant."

Cohen's discussion of the Labor Party's eventual isolation of the revisionists, which led to their withdrawal from the World Zionist Organization and abortive attempts to found an alternative general organization, does not assume that the methods employed were an unqualified success. Compromises that were worked into its ideology in the early 1930s, he asserts, had a great deal to do with political weaknesses effectively challenged, again by the Right, in the late 1970s and into the 1980s.

BYRON D. CANNON

University of Utah
Salt Lake City

D'AGOSTINO, ANTHONY. *Soviet Succession Struggles: Kremlinology and the Russian Question from Lenin to Gorbachev.* Pp. xvi, 274. Winchester, MA: Allen & Unwin, 1987. $45.00.

This is a book written for the Soviet specialist rather than the general reader. That fact, however, does not detract from the contributions of the study.

Since the Revolution in 1917, four major power centers have been in tension in the USSR: the military, the police, the party, and the state bureaucracy. This study deals with the problems of understanding the succession of political leaders in a society struggling with questions of the very nature of the new society.

There is a paucity of materials with which to analyze this so-called system. The internal divisions within a monolithic party, the differences between people and programs on the part of leaders fighting to survive and control, and the nature of the struggle for power in the USSR all contribute to complicate a real understanding of the Soviet Union's succession crises.

This is a tightly written study that uses some newly available material from the correspondence and papers of Leon Trotsky, Boris Souvarine, Boris Nicolaevsky, and other Russian émigrés. This interpretation stresses the programmatic differences between the Leningrad line and the Moscow line—between the emphasis on the workers on the one hand and the peasants on the other; describes the conflict between state communism versus the world revolution; points out differences over an economy open or closed to world trade and investment; and shows how the conflict over one-person rule in a collective governmental structure affected all Soviet politicians.

Each of the Old Bolsheviks—particularly Lenin, Stalin, Brezhnev, and Suslov—emerged from the pack by demonstrating an ability to move from side to side between program differences or, to put it another way, stood above the fray as others took positions leading to their political and sometimes physical destruction.

This centrism, or Machiavellianism, of competing leaders within the Kremlin together with the alternance between programs and personalities may provide a guide to the prediction of future Soviet leadership.

Gorbachev stands heir to this line of succession. By implication, this study would argue that he can survive only as he demonstrates his centrism, his ability to straddle contradictory positions as well as a ruthlessness toward those who would contest for leadership in the Kremlin.

This is an important study.

JACK L. CROSS
Texas A&M University
College Station

ELLWOOD, SHEELAGH M. *Spanish Fascism in the Franco Era: Falange Española de las JONS, 1936-76.* Pp. vii, 207. New York: St. Martin's Press, 1988. $32.50.

POLLACK, BENNY with GRAHAM HUNTER. *The Paradox of Spanish Foreign Policy: Spain's International Relations from Franco to Democracy.* Pp. viii, 196. New York: St. Martin's Press, 1988. $27.50.

Two quite different works are discussed here. The first deals with Francisco Franco's virtually totalitarian helpmate. The second discusses a one-time great power now constrained by forces largely beyond its control. They differ technically in profound degree. The first is carefully—even exhaustively—scholarly in approach but exasperatingly self-restrained in both scope and insight. The second tends toward carelessness and spottiness of documentation and writing and contains little analysis. The constraints of space here force me to be somewhat cavalier in my comments.

Ellwood describes a small nationalist-traditionalist faction of the early 1930s that became the major quasi party of the Franco dictatorship. The civil war (1936-39), together with ferocious opportunism, allowed it to flourish. It was nominally submerged within the official Movimiento National by a decree by Franco in April 1937, but by then its recruitment of both troops and bureaucrats had so built it into the system that it survived until well into the post-Franco republican monarchy. Its peak elite leadership often suffered schisms, but the main organization endured.

Ellwood's book reflects a mix of intentions. She states that the Right has received too little attention in recent literature and the Left

perhaps too much. Yet her product is modest and appears to have lacked professional guidance. The reader receives an essentially two-dimensional and uneven impression of the internal structure and discipline of Falange. A serious criticism is that its ideological positions, and their relevance to the 1930s, are not adequately developed. The discussion gives not the slightest impression of how the party could have survived so long; its very large size and wealth, the penetrative harshness of its hold on power, and the evasive opportunism of its ubiquitous leadership cadres are nowhere mentioned. Her lengthy recounting of intra-elite clashes induces the reader to believe that it was a small sideshow of the most totalitarian national regime the West would exhibit for nearly three decades.

In the final pages, Ellwood surprises the reader, if he or she is otherwise uninformed, by reporting that when Franco died in 1975 the party's estimated annual income from state sources was 9 billion pesetas—about $230 million—and a year later it had a million members in official positions. But in the June 1977 national parliamentary elections, its competing factions received a combined vote of 0.64 percent of the total cast in the country. Here is an unexplained paradox. Ellwood's 11 years of research allowed many original sources to be employed. Surely a more perceptive analysis could have emerged!

The Pollack and Hunter work is an interesting contrast. It is less carefully researched and much less systematically and helpfully documented. It overgeneralizes intermittently and carries the burden of unspecifically defined subjectivity. The book is at its best when it deals with the linkage of domestic Spanish concerns with foreign policy decisions, for Pollack has done previous research in this area. It is much at its worst when dealing with relationships between Spain and the United States. Pollack served in the Chilean Foreign Ministry for several years during the Allende period; it may be possible to see too much in this fact, however.

Spain's historic wish to pursue its own neutralist foreign policy goals is stated often, although the possible utility of this wish is not necessarily clear. For example, its long-standing tilt toward the Arab states conditioned its policy toward the rest of the world for several decades. The reader is presented the possible reasons, but Spain's advantages are not clear. It is suggested that Spain feels some affinity for Third World countries in general, but details of motivation and outcome are not given. Spain's unique role in Latin America's founding is given skimpy attention, and little space is devoted to the subject of its present influence in the region.

The generalized *antiyanquismo* that has led, since the book's writing, to termination of some provisions of treaties with the U.S. Air Force and Navy is described with clearly implied approbation. Given the unmistakable anti-U.S. mood of the writing, one finds disappointingly little substance with respect to causes and events. Much is explained, as it actually is in Spanish contemporary political polemics, as American infringement of Spanish sovereignty, as if a country's sovereignty is forever whole and not subject to conditions. Especially difficult is the fact that Pollack and Hunter demonstrate prejudice by vague and unscholarly language, and employ purposely inadequate and conspiratorial-sounding documentation, when referring to U.S. government policy motivations and actions; this occurs especially in chapter 2. Given the historical facts, this is unsatisfactory.

There are three unfortunate lacunae in the book. First, there is frequent reference to the politicized posture of the Spanish armed forces, but the reality, especially of the period since the death of Franco, is handled with such gentleness and absence of detail that the reader lacks information about the actual importance of the phenomenon. Second, one is told that the Spanish career foreign service has always observed a high degree of professionalism; this gave it an autonomy from the constraints of the fascism of the Franco era and today allows it to influence the leadership of the post-Franco governments. The second fact is so undeveloped that chapter 5, "Making Foreign Policy: The Institutional Framework and the Mechanisms of Democratic Control,"

is simply mistitled. The absence of any summary, analysis, or conclusion is the third lacuna. A pair of generally useful chapters on the European Economic Community and the North Atlantic Treaty Organization terminate the book, but in effect they dangle without integration.

I must conclude, on the whole, that each book has useful factual presentations, but that each is circumscribed by its authors' choices. The second work is the weaker, however, especially since current literature offers competitors of greater scope and better balance.

PHILIP B. TAYLOR, Jr.

University of Houston
Texas

KOLODZIEJ, EDWARD A. *Making and Marketing Arms: The French Experience and Its Implications for the International System.* Pp. xxv, 518. Princeton, NJ: Princeton University Press, 1987. $55.00.

Edward A. Kolodziej presents us with an exceptional book. It is meticulously researched and elegantly written, and it is certainly the definitive work on the linkages between the French armaments industry and French military strategy.

What makes this such a fine volume is that, in addition to the thorough description of the operation of the French armaments industry, Kolodziej explains how the unusual consensus developed for intertwining French political, economic, and military objectives.

Kolodziej's basic argument is that the French armaments industry operates under two imperatives: (1) a very broad-based desire within the French polity to preserve national independence and autonomy of action; and (2) a willingness to use arms both as a stimulant to domestic economic performance and as a brazen tool for gaining or maintaining foreign influence.

One of the most interesting aspects of this volume is the careful historical reconstruction of how the French have seen the role of armaments over time. Kolodziej makes a compelling argument that in the royal, imperial, and republican eras there has been a remarkable consistency in the view about how armaments should be used. It is striking to see that, even under the Blum and Daladier administrations in the 1930s, there was a willingness to use arms production for domestic economic purposes and arms sales for influence.

Although the French have frequently experienced failure in their actual military operations—in 1870, in the early months of World War I, and the disaster of World War II—the French public places a very high premium on sovereignty and freedom of action. As anyone who has dealt with the French in diplomatic settings knows, the French are always so anxious to demonstrate their autonomy that it sometimes appears they take extreme or obstinate positions just to be different.

Kolodziej notes that the Fifth Republic broke sharply with France's imperial past over the desirability of maintaining the colonies, but that De Gaulle forged a coalition of Right and Left opposed to superpower rule:

For De Gaulle, France's capacity to make its own arms was a precondition of a successful challenge to superpower domination. . . . The De Gaulle regime posed this issue in the sharpest and clearest light. Its response—the force de dissuasion—appeared to afford France a measure of control over the overriding issue of peace and war. It also provided the psychological and political bargaining power with which to resist allied demands for alliance conformity and U.S. military and economic expansionism in the 1960s and a counter to rising Soviet military power in the 1970s and 1980s.

These cultural and strategic objectives thus formed the context in which the French arms industry operated. The bulk of this volume then demonstrates how the highly centralized and elite-led French bureaucracy supported and protected its armaments complex. Kolodziej shows how the civil servants in the General Delegation for Armament maintained tight control over both the nationalized and private-sector parts of the industry and how they skillfully kept political support as power

shifted from De Gaulle to Pompidou to Giscard and then to Mitterrand and Chirac in the 1980s.

In France there has also been a willingness to use military research and development and procurement funds to complement or strengthen other economic objectives. Military production facilities are thus widely dispersed throughout the nation, and subsidies and bailouts have been a frequent part of the picture.

The last major piece of this puzzle is French arms export policy. Both De Gaulle and the civil servants running the arms industry realized that it would be impossible to produce a full range of military equipment efficiently if the quantities were just for the French ground, air, and naval forces. Therefore it became necessary to design, build, and market arms that met French needs but that were sufficiently attractive that they could be sold overseas as well.

The French have thus been extremely aggressive arms marketers and have, in a number of cases, been willing to sell to combatants on both sides of regional conflicts. The last part of Kolodziej's book deals with the implications of the "French model" for the global system. Kolodziej is concerned that other middle-sized powers and the newly industrialized countries will imitate the French experience and try to reap the benefits of linking military and technological development strategies.

Here Kolodziej is clearly raising important issues, but we need a good bit more empirical work to be confident of any conclusions. One of the central questions is whether most countries would be able to take advantage of the positive linkages between military and civilian technology. The Brazilians and Chinese clearly think these linkages exist, but it is not clear that countries like Indonesia have been able to benefit from them, and, as the global arms market becomes more competitive, the margins for profit may decline considerably.

In sum, Kolodziej has provided an insightful and authoritative look at a key country and raised very basic questions about how its pattern of development will affect others who try to imitate it.

DAVID B. H. DENOON

New York University

LEWIN, MOSHE. *The Gorbachev Phenomenon.* Pp. xii, 176. Berkeley: University of California Press, 1988. $16.95.

The history of the Soviet Union has been marked by dramatic, leader-inspired turning points, beginning with the 1917 Bolshevik Revolution itself. Lenin's New Economic Policy, Stalin's sanguinary campaign to collectivize agriculture, and Khrushchev's de-Stalinization program all illustrate the pattern of Soviet reform orchestrated from above. The USSR now faces a new and equally dramatic turning point engineered by General Secretary Mikhail Gorbachev.

Gorbachev's campaign to revitalize the Soviet Leviathan is a classic case of history in the making. We are fortunate that a historian of Moshe Lewin's caliber has tackled the challenge. *The Gorbachev Phenomenon* is a significant contribution to understanding the historical context in which the current Soviet reform battles are being waged. Lewin's succinct, vividly written, and thought-provoking analysis will interest the layperson and specialist alike.

It is necessary to state what *The Gorbachev Phenomenon* is not about. It is neither a biography of Mikhail Gorbachev nor a blow-by-blow account of his meteoric rise through the Soviet Communist Party hierarchy. Lewin analyzes a deeper phenomenon—the dynamic, unplanned, and spontaneous social changes that have accompanied the Soviet Union's breakneck transformation from a peasant society in the 1920s to the urbanized, industrial Goliath of today. The rapid pace of Soviet urbanization has spawned a host of social, economic, and political problems. Lewin liberally draws upon recently published Soviet

social science sources to illustrate these various problems. By making these sources accessible to non-Russians, *The Gorbachev Phenomenon* performs an important scholarly function. Lewin's skillful use of Soviet sources also enables us to witness the birth of a critical, probing school of Soviet sociology that is a major ally in Gorbachev's reform campaign.

A striking aspect of the Gorbachev reform movement is its reliance on autonomous social forces from below to supplement Gorbachev's proddings from above. Gorbachev's trenchant critiques of social injustice and bureaucratic stagnation show a shrewd sensitivity to Soviet public opinion. Idealistic yearnings are being reawakened among Soviet youth. Autonomous special-interest organizations are mushrooming among the Soviet populace.

Gorbachev is also taking advantage of structural changes in Soviet society. Rapid urbanization and industrialization since the 1930s have yielded a more highly educated populace. The strategic importance and sheer size of the Soviet intelligentsia are by-products of these dynamic social changes. The USSR's continued existence as a superpower rests on the fecundity of this stratum. It, in turn, requires a free flow of information to remain abreast of scientific and technological developments. Gorbachev's fate may well depend on his ability to turn *glasnost* ("openness") into the battle cry of this increasingly important and self-confident intelligentsia.

Lewin reminds us that these reforms are not written on a blank slate. The New Economic Policy is inspiration for the idea that a socialist state may carefully reduce its commanding role in the economy so as to unleash the productive capacity of the private sector. Similarly, *glasnost* recalls the relative degree of artistic and literary freedom sanctioned by Lenin during the New Economic Policy and which blossomed again during the brief cultural thaw after Stalin's death. Calls for democratization—within the confines of a one-party state, of course—hark back to the Bolshevik party tradition of lively internal debates and factions cut short, initially, by the party's panicky reaction to the 1921 Kronstadt rebellion.

Whither Gorbachev? Ethnic unrest in the USSR, political strife in Eastern Europe, the removal of Gorbachev's ally Boris Yeltsin from the crucial Moscow party-secretary post, and a rancorous literary debate about Stalin's bloody legacy all signify that the Gorbachev drama remains unfinished. *The Gorbachev Phenomenon* is an indispensable guide for those who seek a deeper understanding of both the historical stage and the dramatis personae.

SCOTT NICHOLS

Southern Illinois University
Carbondale

TUGENDHAT, CHRISTOPHER. *Making Sense of Europe.* Pp. 240. New York: Columbia University Press, 1988. $25.00.

LEDEEN, MICHAEL A. *West European Communism and American Foreign Policy.* Pp. viii, 197. New Brunswick, NJ: Transaction Books, 1988. $34.95.

While both of these books deal with European problems, their subject matter is quite different. Their titles are also somewhat deceptive. Tugendhat focuses on the European Economic Community (EEC), usually referred to as the Common Market, while Ledeen deals almost exclusively with the Italian Communist Party and U.S. policy toward Italy since the fall of fascism. He justifies this concentration by the argument that of all the Western communists only the Italian Communist Party threatened throughout more than three decades to enter government and thus undermine Western policy toward the Soviet Union.

Christopher Tugendhat, a former British member of the European Commission, journalist, and member of Parliament, provides a thoughtful and gracefully written account of the history of the EEC, an analysis of what can and cannot be expected from the Community, and a series of recommendations designed to make it more effective. The publication of the book is particularly timely since the Community is currently wrestling with a 1992 deadline for the abolition of all internal barriers to the movement of people and goods.

Tugendhat's wide political experience gives his account of the EEC a unique perspective. His commitment to the success of the Community is clear, but his sober discussion of the problems facing it, such as the perennial fight over agricultural policy and the divisions between north and south, give the reader a clear picture of both the successes achieved by the Community and the realities with which those committed to the European ideal must pay heed. Anyone seeking to understand the EEC should consult this work.

Michael Ledeen's book is clearly an antirevisionist study. He argues that the United States, far from pursuing a policy aimed at helping the Right in Italian politics and thereby seeking to halt change, followed a consistently liberal policy in Italy after the fall of Mussolini. Further, U.S. policymakers, despite some bumbling, especially during the Carter administration, sought to keep the Communists out of the government. In Ledeen's view this was the only rational goal since for all the tactical twists and turns of the Italian Communist Party, it remained Leninist and totalitarian in its structure and aims. For example, he characterizes the once fashionable idea of Eurocommunism as a "myth."

There are some interesting, if controversial, arguments about both Italian communism and U.S. foreign policy in this book. Unfortunately, the text bears evidence of hasty preparation, the prose is often leaden, and the organization of the work—alternating chapters dealing with Italy and the United States—tends to confuse the reader as does the introduction of relatively obscure Italian Communists and radicals into the text without sufficient identification.

FREDERICK J. BREIT

Whitman College
Walla Walla
Washington

UNITED STATES

DENHARDT, KATHRYN G. *The Ethics of Public Service: Resolving Moral Dilemmas in Public Organizations.* Pp. x, 197. Westport, CT: Greenwood Press, 1988. $37.95.

THOMPSON, DENNIS F. *Political Ethics and Public Office.* Pp. viii, 263. Cambridge, MA: Harvard University Press, 1987. $25.00.

The subject of ethics in public life is as timely as the most recent government-contracting scandal or foreign policy debate and as timeless as the disciplines of moral philosophy and public administration, in which these two books are rooted. Although the scope here differs—Denhardt focuses on the practice of ethics in public administration, narrowly defined, while Thompson addresses the ethics of political action in a variety of settings—commonalities remain. The blurring of politics and administration has created the need for administrative ethics, according to Denhardt, and has raised new ethical issues of political action, according to Thompson. Both authors are deeply concerned with the democratic ethos and the U.S. Constitution as a guide for and a constraint upon public officials. Both are critical of our currently limited view of ethical misconduct and its remedies, that is, the emphasis on conflict of interest, financial disclosure, and codes of ethics. Both confront the dilemma of personal responsibility in a system in which many different officials at many organizational levels contribute to policies. Both are explicitly prescriptive in the best tradition of social science.

Denhardt draws extensively upon the literatures of public administration and philosophy in forming her model of ethical administration, which places equal importance on developing ethical public administrators and organizations. As she notes, the obstacles are formidable. Ethical behavior can conflict with professional norms, organizational loyalty, and bureaucratic rules and procedures. Yet a more ethical public administrator, she suggests, can be produced by incorporating more training

in values, ethics, and democratic theory into the professional curriculum. Public bureaucracies, in turn, must develop an organizational conscience, which would both protect the ethical individual and legitimize ethical discourse. She concludes by applying this model in an analysis of a case set in state government.

In contrast, Thompson does not provide an overarching model to integrate his book's seven chapters, most of which have been previously published. This is not to denigrate the insightful and gracefully presented ideas that appear here on topics as diverse as the nondemocratic nature of nuclear deterrence, the culpability of presidential advisers, the moral case for public financing of congressional campaigns, the privacy rights of public officials, the ethics of social experiments, and the paternalism of officials and policies. Throughout, Thompson not only points out the inadequacies of our present approaches to resolving these ethical dilemmas; he also offers provocative alternatives, such as a new criminal-code violation, "the obstruction of the democratic process," to be applied to public officials.

Denhardt appropriately addresses those who teach and practice public administration; the reforms she advocates must come from within. Stylistically more accessible, Thompson speaks not only to scholars but also to the concerned public. Both are realistic about the ambiguities and complexities involved in ethical public action, but both remain resolute, given the importance of these issues.

JANET K. BOLES

Marquette University
Milwaukee
Wisconsin

GERTEIS, LOUIS S. *Mortality and Utility in American Anti-slavery Reform.* Pp. xvi, 263. Chapel Hill: University of North Carolina Press, 1987. $27.50.

By concentrating on the principles underlying American antislavery reform, Louis S.

Gerteis has enriched the literature of this period. If Benthamite utilitarianism, whether in modified form or not, replaced the old values of Christian philanthropy, so political liberalism—as developed by the burgeoning middle class—endeavored to reinterpret old republican values.

Nor was this urge toward liberal values unique in America. As Gerteis explains, Europeans, too, were "viewing themselves as part of a broad struggle for human liberty and celebrating the nineteenth century as an age of progress."

Gerteis also develops important ideas of the role played by the Calvinist tradition in creating the urge for reform. Yet, as is clear from his argument, Calvinist values—expressed here as Yankee Calvinism—were basic to a middle-class ethos. Indeed, the contrast between the hard-headed Yankee Calvinist and the romantic notion, as developed by William Lloyd Garrison, of the release of the slaves "to be free as the winds of heaven" is an intriguing one. It was, however, a different matter if uniting Calvinist piety with individual liberty succeeded in advancing the commercial interests of the North at the expense of those of the—at this time, primarily agrarian— South.

OLIVE CHECKLAND

University of Glasgow
Scotland

JONES, CHARLES O. *The Trusteeship Presidency: Jimmy Carter and the United States Congress.* Pp. xxv, 225. Baton Rouge: Louisiana State University Press, 1988. $24.95.

The Carter administration did not enjoy an especially productive relationship with Congress. Its reputed failures on Capitol Hill have been attributed to the ineptitude and insensitivity of Carter's staff of outsiders. Charles O. Jones, in the first sophisticated scholarly analysis of executive-legislative relations during the Carter years, paints a more

positive picture of the thirty-ninth president's interactions with Congress.

Part of the Carter Presidency Project of the University of Virginia's Miller Center, *The Trusteeship Presidency* relies heavily upon the unique collection of oral histories compiled by a panel of thirty nationally known scholars, including Jones.

Jones sees Carter as explicitly deciding that his would be a "trusteeship presidency." As the newcomer to national politics saw it, it was the president's obligation to represent the public or the national interest. Congress was too concerned with reelection and with special and regional interests to represent the larger national interest. Elected as an antiestablishment candidate, Carter planned to reform and discipline the institutions of government that no longer reflected the public will.

Unfortunately for him, he arrived in Washington during the post-Watergate period when a newly assertive Congress prepared to battle any president who challenged its prerogatives. According to Jones, neither Congress nor the White House understood the way each body was experiencing a period of institutional change.

Carter's trusteeship concept meant less direct consultation with Congress, a return to cabinet government, and a smaller White House staff. Given what Carter set out to do and his model of the presidency, the president's staffing and bureaucratic routines, much assailed at the time, were adequate. Moreover, Jones's box score on legislative victories belies Carter's popular image. His record with Congress was not that much different from the records of previous presidents, beginning with Eisenhower.

Jones's well-written, well-conceived argument is sure to affect debates about the relative success of the Carter presidency. He may, however, accept the testimonies of the president and his aides too much at face value when they claim that unlike the legislators, they were not interested in the relationship between their programs and something as sordid as reelection.

Further, one major problem was Congress's misperception and misunderstanding of Carter's methods and goals. Here, the president may be criticized for failing to explain what he was up to or perhaps to convince Congress of the bona fides of his self-righteous-sounding approach.

All the same, Jones has launched the Carter Presidency Project with an important monograph. His mildly revisionist analysis is likely to be joined by others in the years to come.

MELVIN SMALL

Wayne State University
Detroit
Michigan

LANDAU, SAUL. *The Dangerous Doctrine: National Security and U.S. Foreign Policy.* Pp. xvi, 201. Boulder, CO: Westview Press, 1988. $24.95. Paperbound, $9.95.

America's emergence as a superpower and its cold-war competition with the Soviet Union created what Saul Landau, a senior fellow at the Institute for Policy Studies, calls "the dangerous doctrine" of national security. That is, national security considerations stemming from the Soviet threat were used to justify the making and implementing of vital policies in secret and the use of tactics that Americans would normally judge illegal or immoral. Unfortunately, the results of national security secrecy have not always promoted the national interest. Probably fortunately, more often than not truth has willed out. Acting covertly in the name of national security, American officials in almost all postwar administrations have enacted disastrous policies—for example, the Bay of Pigs invasion and the Iran-contra arms deals—have blatantly misused security considerations to protect their own personal and political interests and to harass domestic critics, as in Watergate and McCarthyism; have distorted information that might have affected major national decisions, as in the Vietnam war; and have overthrown, sometimes bloodily, foreign governments, such as that of Allende in Chile.

The very notion of national security, therefore, deserves some scrutiny if America is to avoid such fiascoes. Certainly, the United States has legitimate security interests; and many nations, both alleged friends and foes, act out of nasty self-interest on the world stage. Also, conspiratorial policymaking can work, as in the case of the Cuban missile crisis. Thus national security is both an unfortunate necessity of the real world and a real threat to American democratic practices and to policy effectiveness. It is certainly a topic worthy of thoughtful analysis that should produce unsettling conclusions.

The dust jacket of *The Dangerous Doctrine* promises such a treatment. The book itself, however, is conceived as a revisionist history and analysis of U.S. foreign policy. Landau argues that several common threads have connected American foreign policy since the founding of the Republic. He sees U.S. foreign policy as primarily driven by economic forces that seek to open up the "land, labor, resources, and markets" of poorer and less powerful nations for exploitation by the American business classes, as America created an "informal overseas empire." The threat of the cold war has melded these economic goals with a strong anticommunism and has been used to justify the growth of a "national security state." While the book presents a comprehensive overview of American diplomatic history, there is a strong geographical focus on Latin America, particularly Central America, based on the implicit assumption that this region is the best example of the "informal empire" that constitutes the essence of U.S. foreign policy.

As a monograph with a controversial interpretation of America's role in the world, *The Dangerous Doctrine* is quite attractive. It develops a coherent theme linked to two centuries of international activities, draws together recent foreign policy highlights, many of which Landau rightly considers lowlights, and—perhaps most important for a potential text—is concise and quite readable. But even many who hold critical perspectives on U.S. foreign policy might consider the interpretive framework a little too neat and unidimensional; for example, is the history of America's retrograde interventions in Central America more of a "sideshow" than an "essence"? In addition, "the dangerous doctrine" itself deserves a more balanced and detailed probing.

CAL CLARK

University of Wyoming
Laramie

LOWE, RICHARD G. and RANDOLPH B. CAMPBELL. *Planters and Plain Folk: Agriculture in Antebellum Texas.* Pp. xvi, 216. Dallas, TX: Southern Methodist University Press, 1987. Distributed by Texas A&M University Press, College Station. $22.50.

This book, in some respects a sequel to the same authors' *Wealth and Power in Antebellum Texas* (1977), is a carefully researched, clearly presented but limited case study of agriculture and economic growth in a frontier region in the last decade of legalized slavery in the Western world. Texas, which did not join the Union until 1845 after its English-speaking population had won independence from Mexico in 1836, was sparsely populated and even more heavily dependent on agriculture than other southern states. The population grew from slightly more than 200,000 in 1850 to slightly more than 600,000 in 1860. The great majority of this population in 1860 had come from other southern states and had brought with them their institutions, their crops and technology, and—some of them—their slaves. Before the Civil War they had effectively settled only the eastern two-fifths of the state, and that is the area with which Lowe and Campbell deal.

They divide this area into four regions based largely on geological and topological criteria, but the regions also prove to have interesting social and economic characteristics. For example, the eastern regions—north and south—had the largest proportions of slave owners, the western regions the largest proportions of landowners without slaves. Overall, the proportion of Texas farmers in 1860 who owned slaves—32.2 percent—was about the same as in Tennessee, Georgia, and Alabama but significantly lower than in Mississippi, where the proportion was 44.8 percent. On the other hand, the proportion of nonslaveholding farmers who owned land was significantly higher—slightly over 50.0 percent—than in the other states.

The principal sources utilized by Lowe and Campbell are samples from the 1850 and 1860 manuscript censuses of the United States, and tax and probate records of the state of Texas. These are supplemented in some instances by records of court cases, a few diaries, and letters and personal records. Although the methodology is quantitative if not cliometric—the book contains 62 tables, several longer than one page—it also contains at least one chapter of narrative history: case studies of six individuals from different regions and different social and economic circumstances. It makes a nice contrast.

The major findings are as follows: (1) slavery in Texas was profitable; (2) production, productivity, and incomes grew rapidly between 1850 and 1860; (3) Texas was essentially self-sufficient in basic food crops; (4) the distribution of wealth and income among the free population was unequal, but less so than in other southern states, and all income groups were prospering with no hint of class struggle. These are concordant with other recent studies and thus not surprising, but they are solidly based and meticulously documented. The bases of the calculations are clearly explained at an elementary level, and Lowe and Campbell are clearly well versed in the recent literature on southern agriculture and slavery.

In summary, this is a model study utilizing both traditional and modern quantitative techniques in which Lowe and Campbell reach conclusions that, although neither novel nor surprising, are reassuringly solid.

RONDO CAMERON

Emory University
Atlanta
Georgia

PINDERHUGHES, DIANE M. *Race and Ethnicity in Chicago Politics: A Reexamination of Pluralist Theory.* Pp. xix, 318. Champaign: University of Illinois Press, 1987. $29.95.

Richard Daley, unlike most recent local politicians, overwhelmingly dominated Chicago politics for over 20 years from at least the time when he became mayor in 1955 to his death in 1976. This is not clearly evident from a reading of Dianne Pinderhughes's recent book on twentieth-century Chicago politics. To be sure, after briefly discussing Mayor Harold Washington's campaign and election, Pinderhughes begins with Mayor Daley. She identifies him as a "formidably talented political leader. . . . oddly charismatic." At the close of the book, there is a brief discussion of Daley's death and the politics of the post-Daley era. Though labeling the Daley era as the height of machine power, the book is mostly an examination of the historical roots of "ethnic politics," which means the years from about 1910 to the 1930s. Mayors Big Bill Thompson and Anton Cermak of that era rate comparatively more space than Daley and his successors.

It is the pluralist notions—mostly from Robert Dahl and James Q. Wilson—of eventual ethnic, and especially racial, integration into American politics that most concern Pinderhughes. To Pinderhughes, Chicago politically was not and has not been open and competitive but discriminatory and racist. Specifically, she focuses on the political experiences of two white ethnic groups—Poles and Italians—as well as blacks. The Poles and especially the Italians were not systematically incorporated into Chicago government, in part because they were relatively small, dispersed, or church focused. But it was blacks who were excluded from major city hall decisions as "ethnic and especially racial discrimination dominated life." This may be changing, however, with the assumption of power by Mayor Washington, who was a black. Unfortunately, even though the book is dated 1987, there is little about the Washington or the post-Washington era.

In two rather carefully crafted chapters on crime and education, Pinderhughes contrasts the three groups during the early twentieth century, with many figures provided. In each case, she asks a number of probing questions. For example, "Did the police arrest and prosecute [blacks] more frequently [than others]?" To which, she answers, "Whites and blacks committed crimes; blacks were punished for them." Other chapters, however, are less clearly focused. For instance, one is entitled "The Philosophical Labyrinth of Race and Political Beliefs," which ranges all over the country and the issue. Probably it would be better placed elsewhere.

If one is looking for a book that examines big-city politics during the early years of the twentieth century, then this one might fill the need. But it should be noted that it has a number of relatively minor flaws. One is the lack of supporting data for many generalizations. For example, Pinderhughes asserts that "black voting is not now nor has it consistently been low." What is the evidence? Also, what is the evidence that the "[Italian] family's control over women seems to have increased" or that "blacks employed in organized crime . . . were . . . much more frequently prosecuted"? How did "[Byrne] repeatedly remind . . . blacks of their subordinate positions within the machine"?

The book has some unclear statements and grammatical errors. For example, "there is data" should be "there are . . ."; "neither Polish or" should be ". . . nor"; and on page 46 "political" is misspelled. More fundamentally, "descriptive" and "substantive" representation and "tangible . . . and substantive aspects" should be more carefully distinguished. Did the percentage of blacks shrink in the second ward between 1930 and 1940? Was Powers alderman of both the nineteenth and the twenty-fifth wards? Are "four in the house, [and] four . . . elected state representatives" not the same? Did Democratic Mayor Dever appoint a Republican? How can an interview of 11 March 1974 be evidence that "between 1910 and 1980 only four blacks served on the county commission"? Pinderhughes describes Daley as "one of the few living politicians," but was Daley not already dead when she wrote this book? Who alludes to whom in "these coalitions allude to James Q. Wilson's analysis"? What is "imprinted them"?

There are many of these rather off-putting errors and unsupported assertions. Also, the book will probably please neither the scholar of the theoretical debate nor those interested in the dominating Richard Daley. Yet, on balance, it provides much insight into the history and politics of one important city. For the student of the Chicago scene, as well as urban and racial politics in general, it is a must.

WILLIAM M. BRIDGELAND
Michigan State University
East Lansing

REINSCH, J. LEONARD. *Getting Elected: From Radio and Roosevelt to Television and Reagan.* Pp. xiv, 338. New York: Hippocrene Books, 1988. $18.95.

TEIXEIRA, RUY A. *Why Americans Don't Vote: Turnout Decline in the United States, 1960-1984.* Pp. xvii, 348. Westport, CT: Greenwood Press, 1987. $55.00.

J. Leonard Reinsch, retired chairman of Cox Broadcasting, became involved in politics in 1944 when Franklin Roosevelt asked James Cox to loan him a "radioman." Reinsch served the Democrats running their national conventions and planning the use of electronic media through 1968. Although discarded by the McGovernites in 1972, he continues his descriptive analyses of both parties' conventions and campaigns through 1984 and projects some facets of 1988.

Political scientists and participants alike have viewed with alarm the deterioration of political parties and the increasing costs of politics. Reinsch elaborates these concerns and advocates changes in both the process and financing. A new election day—the second Sunday in November—with concurrent 12-hour voting periods across the country reflecting time differentials, coupled with nationwide

registration by postcard, is proposed. To improve the nominating process and reduce costs, Reinsch would divide the country into four regions with a uniform caucus or primary date for each region and a one-month breather between each vote.

Operational changes for national conventions are also proposed but the most radical suggestion is to place the vice-presidential nomination in the hands of the newly elected National Committee at a postconvention meeting. Shorter campaign periods and a limit on television expenditures would save money.

Reinsch, who established the television debate for Kennedy, believes such debates are a given for the future since three incumbent presidents have legitimated the process. He concludes that cable and videocassette recorders will supplant television in future elections.

Ruy Teixeira, a sociologist serving as a senior analyst specializing in survey research for Abt Associates, provides a valuable addition to the current studies of nonvoting in America. His data set, derived from the American National Election Studies, has been the basis for other comparable studies. Teixeira's analysis focuses on the Abramson and Aldrich work (1982) since the same time span is covered. Abramson and Aldrich are faulted for confining their data to whites and for omitting education, occupation, and income variables from their model. In addition, three variables that support decline—enfranchisement of 18-year-olds, increase in residential mobility, and increases in the proportion of those not living as married couples—were not included in their data.

Teixeira's own analysis indicates that inclusion of age and life-style variables would help explain the decline in turnout. At the same time, America became more middle class from 1960 to 1980, which should have increased participation. Teixeira concludes, "Within the context created by this SES upgrading, it becomes apparent that factors such as age, newspaper reading, marital status and mobility must be considered in addition to partisanship and efficacy if the fall in turnout is to be adequately explained."

When the time span is divided into two segments, Teixeira finds that the 1960-68 period decline is overwhelmingly related to political efficacy while the decline in newspaper readership is the crucial development for 1968-80. Yet both periods were times of socioeconomic upgrading, which should have improved voter turnout.

Because turnout improved in 1984, Teixeira measures his model by that election. His significant factors all support an increase in participation but at a higher rate than actually occurred. Consequently, an "unknown factor or factors"—also potentially evident from 1960 to 1980—appear to produce a counterforce. The critical realignment in American politics, which should have occurred cyclically around 1968, still has not happened. Instead, Teixeira senses that a "dealignment" has taken place, and he concludes that voter turnout will likely increase again only if there is a realignment or the installation of a collectively oriented system of voter participation that relieves individual responsibility for voting.

While these two books represent very different approaches to American politics— one coming from a descriptive practitioner and the other from a statistical behavioralist—I sense that the authors of both feel that the impact of television and the decline of political parties are critical factors in the changing nature of our elections. Both books belong in serious library collections.

J. H. BINDLEY

Wittenberg University
Springfield
Ohio

ROBERTS, ROBERT N. *White House Ethics: The History of the Politics of Conflict of Interest Regulation*. Pp. 215. Westport, CT: Greenwood Press, 1988. $37.95.

Robert N. Roberts's *White House Ethics* is a readable, straightforward survey of the legislative efforts to curb honest graft in the administrative branch. Organized chronologically, the book quickly covers the years before World War II, which Roberts sees as a turning point; a chapter is then devoted to each of the postwar presidencies. Roberts has no ideological axes to grind, so his careful account will provide useful background information for anyone interested in gaining a perspective on the scandals of the Reagan era.

World War II vastly increased the discretionary power of government and hence, Roberts argues plausibly, the potential for executive branch abuse. While the New Deal was notably free of conflict-of-interest scandals, the Truman presidency was hurt by both the waning of the New Deal ethic and the Missourian's tendency to rely on his old cronies, "the five per centers." The shift was symbolized in the person of Tommy "the Cork" Corcoran, a one-time New Deal kingpin who became the Truman era's fixer par excellence.

The narrative builds to the 1978 Ethics in Government Act, which was the culmination of post-Watergate efforts at executive branch ethics reform. The reform attempt failed to halt influence peddling, Roberts argues, because the enforcement mechanisms were always the subject of partisan wrangling. Worse yet, the complicated procedures put in place by the 1978 Carter reforms tended to discourage Washington outsiders from seeking federal office. The result was that from 1978 on, executive branch positions were increasingly filled with resident ideologues from the Washington think tanks.

Useful though it is, *White House Ethics* would have been a far richer book if it had looked, at least in passing, at the ethical issues of the legislative branch. The influence peddling that convicted Reagan aide Mike Deaver under the 1978 reform law would have been perfectly legal for a congressional aide. Similarly, it would have been useful if Roberts had considered the impact of James Buchanan and public-choice theory on attitudes toward governmental ethics. If Buchanan is right and government officials, no less than entrepreneurs, are rent-seeking income maximizers, then conventional attempts to deal with governmental ethics are bound to come to naught.

FRED SIEGEL

Cooper Union
New York City

ROSSWURM, STEVEN. *Arms, Country, and Class: The Philadelphia Militia and "Lower Sort" during the American Revolution, 1775-1783.* Pp. xv, 373. New Brunswick, NJ: Rutgers University Press, 1987. $40.00.

In his introduction, Steven Rosswurm tells his readers that, as a Marxist historian, he begins "with the premise that the production and reproduction of human existence and the class relations through which that production and reproduction are organized and the surplus is appropriated establish the primary 'boundaries of human existence.'" Thus Rosswurm stresses economic goals as motivation for human actions and unequal class relations as cause for human controversy.

The book focuses on the Philadelphia lower sort from 1775 to 1783 and tells how their participation in the militia during the Revolution allowed them to gain power temporarily. Rosswurm reports that by 1775, the city's laboring poor were insecure, subordinate, and dependent upon the top 10 percent of the population who owned 73 percent of Philadelphia's taxable wealth. After the battles of Lexington and Concord, the people began to form a militia, whose majority soon consisted of poor laborers. Their demands for equitable militia regulations caused them to become politicized. Although they never secured equality of treatment, they were able, with leadership from an elected Committee of Privates, to exert pressure on the upper classes and to influence some decisions. For example, a coalition of militiamen with middle-class artisans forced Pennsylvania into the independence movement, ended the proprietary government, and backed a very democratic new state constitution. Their ability to influence affairs ceased, however, after 4

October 1779. That day, gunfire broke out between men in James Wilson's house and militiamen on the street, who were marching to protest wartime inflation, monetary depreciation, inadequate militia compensation, speculator manipulations, and the continued presence of tories. After "Fort Wilson," the laboring poor "slipped back into virtually their pre-1775 state of powerlessness," according to Rosswurm.

Rosswurm writes that "a social history devoid of politics and the context of class relations is flawed history." One might also argue that a social history that is exclusively devoted to politics and class relations is also lacking. For example, the narrowing of discussion to economic matters neglects the importance of religion in a colony where religion was very important. Not all the poor joined the militia. Many Quaker poor did not, yet they had the same economic motivation as those who did. To ignore their peace testimony is to create an incomplete picture.

Within the limits set in his introduction, however, Rosswurm has presented an interesting and well-researched picture of how the Philadelphia lower sort who joined the militia affected and were affected by the course of the Revolution.

ANNE M. OUSTERHOUT

Michigan State University
East Lansing

SMITH, J. OWENS. *The Politics of Racial Inequality: A Systematic Comparative Macro-Analysis from the Colonial Period to 1970.* Pp. xv, 202. Westport, CT: Greenwood Press, 1987. $37.95.

BUMILLER, KRISTIN. *The Civil Rights Society: The Social Construction of Victims.* Pp. x, 161. Baltimore, MD: Johns Hopkins University Press, 1988. $19.95.

These valuable books represent two very different approaches to the study of race in the United States. The Smith book is a comparative historical analysis of racial inequality.

Drawing on political economic data, Smith explores the dynamics of stratification and mobility as encountered by different ethnic groups at different periods in the nation's history. The Bumiller book is a critique of legal ideology and discourse as they operate in the field of civil rights. Drawing on a small sample of intensive interviews, and heavily influenced by Foucault, Bumiller argues that law is an ineffective vehicle for challenging discrimination or promoting equality.

Social-scientific analysis of race and ethnicity is being reconstituted today. Thus, disparate as these two works may initially seem, there is an underlying unity in their approaches. Both question the established wisdom about their subject. Both are forced to reconceptualize by the limits of the available theoretical and historical accounts. Smith challenges the theoretical logic displayed in the vast majority of comparative ethnicity-based approaches to race. As I have argued elsewhere, ethnicity theory displays a near-universal tendency to lump together the most variegated experiences of minority groups. For example, writers as diverse as Glazer and Moynihan, Sowell, and Wilson have all argued that the fit between minority and core group values determines the degree of upward mobility available to the minority group in question. Because they argued this across the board for all ethnic and racial minorities, as well as historically, this homogenizing approach became the centerpiece of their general theories of U.S. ethnic and racial dynamics.

Smith effectively challenges these approaches by disaggregating data on income inequality by different ethnic and racial minorities and by different historical periods. As a result we learn a good deal about the different routes to upward mobility available—and not available—for the groups in question. I am convinced by his argument that "to make it to the middle-class plateau collectively, a group needs two things: a pair of boots and a set of bootstraps." In this formulation, the boots are the resources available to the members of the group themselves, namely, human capital.

The bootstraps are the systems by which the state legally protects minority groups' economic opportunities. Not surprisingly, the state turns out to have played a crucial role in affording or limiting opportunities. Smith argues that at crucial moments the state provided bootstraps to ethnic and racial minorities through such mechanisms as the Homestead Act of 1862, the National Labor Relations Act of 1935, and the Civil Rights Act of 1964.

The question unavoidably arises of how effectively civil rights legislation in the 1960s provided bootstraps for blacks. Although Smith states on his final page that these laws did serve this function, he seems unconvinced that blacks have achieved the same mobility-generating combination as did other, non-racialized ethnic groups. If indeed they have not, this suggests some limits to Smith's approach. Although he certainly aims to do so, I wonder if his comparative methodology sufficiently distinguishes between racialized and nonracialized groups. Smith may underestimate the political importance of racism by confining his application of this category to macro-level, public policy arenas.

On this question, Bumiller's approach leans to the opposite extreme. In her analysis, law is an ineffective tool to counter discrimination, racial or otherwise, because legal discourse and practice cannot address the real dynamics of power. Law creates identities, constructs victims, restructures power relations, but, according to Bumiller, always in an ideological, almost fictive way. It inevitably perpetuates a political illusion by suggesting that legal reform will facilitate a transformation—perhaps a gradual one—of society, fostering equality and eliminating discrimination. But based on wide-ranging interviews with a small sample of "victims" of discrimination, Bumiller attacks this as an "illusion of individual rights" ideologically encoded in law. In her view, "law is incompatible with everyday life." Discrimination operates quite effectively beneath the law. It is apparent at the level of subjectivity, as a component of identities and power relations; mere legal reform cannot affect this reality. In fact, Bumiller suggests that civil rights legislation may paradoxically exacerbate discrimination by rendering continuing perceptions of inequality less legitimate in the eyes of the "victims," thus making mobilization against discrimination more difficult.

While this view contains an important insight, it fails to appreciate the complexity of contemporary U.S. politics. Just as Smith limits his case by looking at politics only as a way of achieving economic gains, Bumiller limits her argument by writing off the social-structural features of antiracist—and anti-sexist, and so on—politics as mere illusions. This is more than apparent in her extremely brief treatment of the civil rights movement—spanning three pages—in her glib conceptualization of the public-private distinction, and in her assertion that the civil rights movement merely "has the appearance of challenging the accepted boundaries between social, economic, and political life." It is difficult to believe that Bumiller has proven this assertion. Certainly her interviews do not demonstrate or even seriously address it, and there is a great deal of analysis—most of it not considered by her—that suggests the opposite. Yet she is correct to stress the importance of identity, of the way that politics frames "the reality of personal lives."

Both Smith and Bumiller would do well to consider how the politics of identity, of movements, transformed the everyday life of blacks and other minorities during the dramatic struggles of the 1950s and 1960s and what the legacy of this transformation has been. The long-term gains made then, though not without their problems and reversals, transformed the meaning of race and the nature of political identity in general, such that not only economic and political equality but also social and cultural equality were placed permanently on the political agenda. From that intervention we have derived all the new social movements, and indeed the theoretical frameworks upon which Smith and Bumiller draw. Since those

political labors were undertaken, the debate has been not about whether but instead about how equality is to be achieved, and indeed what equality means.

HOWARD WINANT

Temple University
Philadelphia
Pennsylvania

SOCIOLOGY

BENNETT, GEORGETTE. *Crimewarps: The Future of Crime in America*. Pp. xix, 435. Garden City, NY: Doubleday, Anchor Press, 1987. $19.95.

Crime is alive and well in the United States, but the composition of criminals and victims is likely to change—definitely but gradually—in the next two decades. Essentially, that is the message of *Crimewarps*, written by one who is a mixture of sociologist, consultant on crime prevention programs, and a media spokesperson. Written in a pungent, *Newsweek*-like style, with many vignettes and rather crisp chapter summaries—but without the encumbrance of statistical tables and graphs—this is a book for a wide audience.

The neologism "crimewarps" refers to the presumed redirection or bending of recent trends in the entire criminal realm. The book is therefore an analysis of imminent shifts in the parameters of crime, based on Bennett's reading of various statistical barometers. She concludes that, in the near future, criminals will be older, more frequently female, more middle class or white collar, more suburban and rural in location, more regionally dispersed, and somewhat more involved in so-called victimless, moral types of crimes. But she also predicts somewhat lesser use of drugs and more effective drug control—a prediction that on her own evidence, still seems quite optimistic. Finally, she predicts that public handling of crime will inevitably infringe on privacy and civil liberties and that both the definition and treatment of crime and criminals are part of a larger moral and political conflict in our nation.

Crimewarps offers a needed correction both to the law-and-order exponents and to those who retain a liberal-rehabilitative approach to criminals. Instead, it focuses not on causes but on the basic but remote factors—demographic patterns, economic changes, and technological and geographical opportunity structures—that help in interpreting the multitude of rates in a time slice of criminal activities. Though it has been professionally noted and studied for over four decades, the range and significance of white-collar crime is rightly emphasized in this volume. Perhaps not enough attention is given to the enormous variety of crimes and criminals in modern society and thus to the complex causal chains through which sets of individuals fail to learn normal modes of social behavior or through which they deviate from what they may have previously learned. Nor is there much concern for the pivotal upsurge of crime following World War II in most Western nations. But *Crimewarps* compels us to regard crimes—like death and taxes—as disturbingly normal accompaniments of human societies, which continually pursue adaptation in contradictory ways.

ALVIN BOSKOFF

Emory University
Atlanta
Georgia

DANIELS, NORMAN. *Am I My Parents' Keeper? An Essay on Justice between the Young and the Old*. Pp. 194. New York: Oxford University Press, 1988. $19.95.

RIVLIN, ALICE M. and JOSHUA M. WIENER with RAYMOND J. HANLEY and DENISE A. SPENCE. *Caring for the Disabled Elderly: Who Will Pay?* Pp. xviii, 318. Washington, DC: Brookings Institution, 1988. $29.95.

The elderly command a compelling attention in contemporary America: the sheer number of the aged—defined as the elderly aged 65 and over—is staggering. Estimates of 31.3 million aged 65 and over in the period 1986-90 will by 2016-20 have increased to 50.3 million.

The social problems posed by so massive an evolving demographic shift are, candidly, overwhelming, not only in the area of long-term medical care for the elderly but also in the more elusive less definable ethical issues surrounding the biomedical realities and social obligations that accompany these changes.

The two works under review address the issue of the elderly in neatly packaged patterns. Daniels, of Tufts University, quite correctly—as a professional philosopher—subtitles his slender volume "An Essay on Justice between the Young and the Old," and his essay is a carefully reasoned disquisition in biomedical ethics. Daniels's major inquiry is, in his words, "about the just distribution of resources between the young and the old," and he "seeks a principal way, rooted in a theory of justice, to resolve disputes about how income support, health care, and other social resources should be allocated to different age groups in our society." Primarily concerned "with the just design of social institutions," Daniels's essay is a philosophical inquiry and is only incidentally concerned with analysis of policy options.

Rivlin and her associates at the Brookings Institution, in contrast to Daniels, are precisely concerned with policy options in providing long-term care for the elderly, and their work, under the auspices of the distinguished Brookings Advisory Panel on Long-Term Care, is a detailed analysis of the major options for reforming the way long-term care is financed.

Reading Daniels's essay first provides one of many ethical bases from which the intricate social policy formulations of the Brookings Institution investigators may be both understood and assessed. In a truly felicitous way, the works of Daniels and Rivlin and her associates are complementary and afford invaluable, if troublesome, philosophical and ethical insights into, and a pragmatic management of, a grave contemporary social problem of compelling moral importance.

At best, there can be little agreement on a philosophical bedrock on which public policies are to be structured. Daniels, apparently a secular humanist, does not draw in any substantive way from Hebraic-Christian ethics— a vast, complex context of ritual and practice strengthened by a millennial evolution in addressing human needs; instead, he argues that since all people pass through institutions that distribute social goods over the whole life span, each of us can determine what it would be prudent to give ourselves at each stage of life, and this very process—"the Prudential Lifespan Account"—would allow us to discover what justice requires between age groups. He argues:

I offer a unifying vision. We all pass through institutions that distribute goods over our lifespan. If these institutions are prudently designed, we *each* benefit throughout our lives. It is only prudent to treat ourselves differently at different stages of life, as our needs change. What is prudent with respect to different stages of a life determines what is fair between age groups. Prudence here guides justice. If as policy makers, planners, and the general public we can all keep our eyes on this unifying vision, and if we can ignore the divisive talk about competition, then our target will be policies that benefit us all over our whole lives. Establishing such policies would mean doing justice to the old and the young (p. 155).

This argument is unconvincing; worse, given human nature, it is probably impracticable.

Rivlin and her associates are untroubled by Daniels's concerns; hard-headed social pragmatists, their concerns in caring for the disabled elderly pose questions for carefully constructed answers: (1) is there a better way to organize the delivery system? (2) is there a better way to finance care? (3) how should public and private responsibilities be divided? Their answers are meticulously crafted. Contrary to Daniels, who argues that shifting the responsibility, in part or whole, of caring for the elderly from public budgets to family budgets would intensify the struggle for resources between the elderly and the young, the

Brookings investigators devise a significant place among their recommendations for family care. They delineate the family role in long-term care and construct paradigms for long-term block grants and long-term care insurance.

The Rivlin monograph is based on two premises: (1) rising costs of long-term care should not come as an unpleasant surprise; rather, chronic disability is a normal risk of growing old that can be anticipated and planned for, both publicly and privately; and (2) neither the public nor the private sector can handle these rising costs alone. Out of their painstaking examinations come a series of conclusions that shape their recommendations, and these are convincingly made: long-term care is not covered to any significant extent either by Medicare or by private insurance; risk pooling is appropriate to long-term care financing; both public and private efforts are needed to finance long-term care; the primary source of public sector financing for long-term care should be a social insurance program rather than a welfare program; and private and public sector financing can fit together in a variety of ways to cover long-term care. How compelling and thoughtful their analyses are is illustrated by a brief excerpt:

The best solution to the long-term care problem would be medical breakthroughs that reduced the prevalence of disability among the aged and diminished their need for care. Both public and private funding should be directed to biomedical research to reduce the incidence of disabling diseases and to develop techniques for managing them better. However, success is not assured. Advances in medical knowledge are inherently unpredictable and can have the effect of postponing death while lengthening periods of disability. Even with rapid medical advances, it is likely that the number of disabled elderly will increase for many decades to come. It is also likely that both private and public expenditures will rise rapidly and that the fraction of long-term care patients dependent on welfare will not shrink. Public policy must be predicated on these assumptions (pp. 12-13).

Both the Daniels and Rivlin books are important contributions. Admittedly, Daniels has the more difficult task, and it might be remembered that philosophical consolation for old age is always risky even if argued in a modern social idiom. Cicero's *De Senectute* encountered only some small success a long time ago. The social engineering of the Brookings investigators remains a more palatable empirical model, essentially impersonalized, but forthright and convincing.

FRANCESCO CORDASCO
Montclair State College
Upper Montclair
New Jersey

GIBSON, MARGARET A. *Accommodation without Assimilation: Sikh Immigrants in an American High School.* Pp. xii, 244. Ithaca, NY: Cornell University Press, 1988. $34.95. Paperbound, $12.95.

In the growing literature on Asian immigrants to the United States, the present volume is of particular interest. Unlike the highly educated Asian Indians who have settled in metropolitan areas since 1965, the immigrants who are the subject of this volume are farmers, mostly Jat by caste and Sikh by religion, who came directly from the villages of Punjab and settled in the rural Sacramento valley in a town Gibson calls "Valleyside." Punjabi farmers had settled in California at the turn of the century, but this was stopped by legislation in the 1920s. Immigration resumed after 1965, and by 1981 over 6000 Punjabis had settled in the valley.

With little previous contact with Western life-styles, the Punjabis "arrive with few salable skills, speak little English, if any, and, in the case of the women in particular, have had little formal education." Yet theirs is an immigrant success story. Beginning in grueling, underpaid jobs as farm laborers, they are usually able to buy their own home after 5 years, and after 10 years or more, they have become small farmers owning half the peach acreage in the area, some of the most productive orchard land in California. Though three-quarters of the

Punjabis are still laborers or factory workers and the average family income is considerably less than that of the non-Hispanic Valleysiders, their children earn better grades in school and are more likely to graduate from high school and go on to college.

That these achievements are the product of persistence and hard work, and that the Punjabis encounter a great deal of prejudice along the way, should come as no surprise. Margaret Gibson was encouraged to conduct the present study in the early 1980s by local educators who were concerned about ethnic tensions in the school and the community. Her team collected data on the school as a whole and held intensive interviews with the 42 Punjabi families with a child in the senior class and an equal number of Valleysider families.

Gibson delineates the cultural, attitudinal, and situational factors involved. In some respects, the fit is good: for Punjabis, America is the land of opportunity. They appreciate its openness, and their own self-confidence, egalitarianism, and assertiveness as Sikhs equip them to take advantage of it. Also, they come to America through family networks, a route facilitated by provisions of the current immigration law. In other respects, they have made major adjustments—the "accommodation" of the title. In America, they do back-breaking menial labor that would be demeaning for them in Punjab, their women work outside the home, and they place a far greater emphasis on formal schooling, for daughters as well as sons, than before. They have also had to learn to endure the ethnic insults they encounter. Finally, some traits ensure their ethnic "enclavement," notably their extended family system and deeply held convictions about family honor, especially as regards the behavior of their daughters. They help their relatives and resist interactions with the majority that are perceived as impinging on their family values. In school, most of them study in special sections for those who have English as a second language, and some have severe language problems. Overall, they perform better than the Anglo majority partly by default: the majority place far greater emphasis on extracurricular activities than on schoolwork, activities in which the Punjabis are unable or unwilling to join.

Gibson's discussion of the attitudes and myths of the Valleysider majority is less detailed. She makes no specific policy recommendations but implies that the majority would do well to accept "accommodation without assimilation" as legitimate instead of regarding it as un-American. She discusses her findings in the light of various theories and by comparison with other immigrant groups in the United States and Britain. Her style is quite readable, and her contribution will be valuable to educators, South Asianists, and students of ethnic relations alike.

LELAH DUSHKIN

Kansas State University
Manhattan

GOYDER, JOHN. *The Silent Minority: Non-respondents on Sample Surveys.* Pp. vii, 232. Boulder, CO: Westview Press, 1988. $39.00.

Sociologists rely heavily on survey results in developing their theories of society. It would probably be fair to say that census and survey results are used in over 90 percent of the recent articles in major journals. The problem of nonresponse to surveys is therefore important to all persons engaged in social research. The stated goal of the book is to develop a theory of nonresponse. Surveys from Canada, England, and the United States are used.

Goyder contends that studies on the sociodemographic correlates of response behavior are a massive but poorly integrated literature. Goyder reviews previous findings and conducts a number of regressions using previously published results as cases. In general, response rates in survey research fell during the 1960s, but not after that. Response rates are highest in the United States. Controlling for salience of topic and other variables, British surveys

with a response rate of 70 percent would have been 76 percent in the United States. About half of the population would strongly object to detailed inquiries about money matters, but only 6 percent would object to questions about job or occupation.

The Silent Minority is a good review of the literature on survey nonresponse. The book also is good at illustrating the problems of survey researchers. But attempts to develop a theory of nonresponse fail. Goyder notes the fact that response rates of mailed surveys go up if a stamp is provided, and then he attempts to relate this to exchange theory, which weighs costs and benefits to the responder and researcher. But Goyder then drifts off into a history of exchange theory, discussing how cross-cousin marriage is embedded in tribal culture and how Jacob toiled seven years for Laban. I doubt the discussion would be found interesting or germane by any likely reader of this book. In fact, the discussion of cross-cousin marriage, judged in itself, is superficial and unlikely to be of interest to anyone with detailed knowledge of such systems.

The Silent Minority can be recommended to students and faculty looking for a good survey of the literature. The secondary analysis of previous studies is interesting, with Goyder's methodology clearly explained. But do not look here for any good theoretical reason as to why people do not respond to surveys. There may be no good theory, though Goyder does confirm that people do not like to respond to questions about their incomes or other touchy subjects.

GEORGE H. CONKLIN

North Carolina
Central University
Durham

JOYCE, ED. *Prime Times, Bad Times.* Pp. 561. Garden City, NY: Doubleday, 1988. $19.95.

BOYER, PETER J. *Who Killed CBS? The Undoing of America's Number One News*

Network. Pp. vi, 361. New York: Random House, 1988. $18.95.

Since the World War II days of Edward R. Murrow, CBS radio and, later, television news has been the standard for broadcast journalism. Murrow and the journalists he recruited were tough-minded reporters, able both to reach the hearts and minds of their audience skillfully and to marshal in a coherent fashion the relevant facts of the most important stories of their times. Eventually, Murrow was to engage in a losing confrontation with CBS management centered about his view of the objectivity and purity of journalism versus the overriding concerns of the corporation that financed the news-gathering apparatus of CBS.

These two books bring the story of the Murrow legacy up to date, and they show in convincing detail the overwhelming defeat of the journalistic purists. As befits the television correspondent of the *New York Times*, Peter Boyer writes an intelligent, objective, and perceptive account. Ed Joyce was once the president of CBS News. He was a prime player in the changing of standards at CBS but was later to be fired as part of the political struggle that was occurring there. Accordingly, his book is full of detail and is in some measure a very human attempt to vindicate his role and his reputation as a journalist.

The events of several key dates vividly illustrate the change at CBS. One was Valentine's Day, 1980, when Dan Rather was announced as the replacement for television anchor Walter Cronkite; Rather's opposition had been the scholarly and thoughtful Roger Mudd. The selection itself was not as important as the reasons for the selection. Rather was and is a first-rate television journalist, but he was chosen because he was a more dramatic journalist than Mudd and because his agent had played an excellent bargaining game with CBS management. The journalist as super-celebrity was supreme.

In late 1981 Van Gordon Sauter was named CBS News president, and Ed Joyce was installed as his deputy. Sauter was brought in not because of his news background but rather because of his flair for the dramatic. Joyce was to be given the assignment of budget cutter and was to make certain that the news division adhered to tough, new financial guidelines.

Standards had previously been to get the story, get it right, and get it at any cost. The new motto was to get the story inexpensively but tell it with a flourish. Cults of personality began to arise at CBS, and symbols of the old regime, including even Cronkite, were downgraded.

The process continued for several years, and even with the departures of Joyce and Sauter, the old CBS journalistic standards seem not to have returned. Many of the CBS correspondents are excellent journalists—political analyst Bruce Morton and Charles Kuralt, among others—and rank at the top of their profession; the CBS news stable is quite possibly still the best in broadcast journalism. But these outstanding reporters no longer seem to have a champion arguing their cause in the highest echelons of CBS. This is a loss for CBS News and for a nation that needs every possible source of objective information about the world in which we live.

FRED ROTONDARO

Congressional Affairs Press
Alexandria
Virginia

WUTHNOW, ROBERT. *The Restructuring of American Religion: Society and Faith since World War II.* Pp. xiv, 374. Princeton, NJ: Princeton University Press, 1988. $25.00.

No one book could possibly do service as a sociological map of the current state of American religion, but if a reader were confined to only one, this could, arguably, be it. To be sure, Robert Wuthnow, a distinguished sociologist at Princeton University specializing in religion, leaves some gaping lacunae in the map: nothing on the black churches; minimal attention to Mormonism and the American spiritualist traditions; and only slight and entirely statistical focus on Catholicism, which is too easily assimilated in its peculiar and radical restructuring to the pattern of mainline religion. Indeed, in many ways, the book could have been more appropriately entitled *The Restructuring of American Protestant Religion.*

As a graduate student at the University of California, Berkeley, Wuthnow apprenticed with both Charles Glock, the best survey-research student of religion of his generation, and Robert Bellah, an innovating theorist who links sociology of religion, in the classic way, with larger social theoretical concerns. Wuthnow combines here the very best of both his mentors' strengths and styles.

By appealing to both quantitative measures and more qualitative content analysis of religious discourse, Wuthnow presents an ideal type of American religion prior to World War II and through the 1950s. He contends that a major restructuring of American religion has occurred since that time. American religion was remolded by the force of changes in the larger society: increased education, geographic and social mobility, technology. Indeed, education turns out to be the most important single predictor of religious attitudes and values. With increased levels of education, "liberal" religion in America has also significantly increased, but not the mainline Protestant denominations.

Wuthnow points to five major restructurings. The first is the declining significance of denominationalism. An earlier period saw denominationalism—or, at least, Will Herberg's larger tripartite Protestant-Catholic-Jew—as a significant American focus of identity, commitment, loyalty, and social interaction. But, as H. Richard Niebuhr argued sixty years ago, American denominationalism was rooted in social factors. With time, the classic interdenominational cleavages and deep conflicts have diminished. Church switching has risen from 4 percent in the 1950s to nearly 33 percent in the 1980s. Intermarriage rates across denominational divides have soared. Members of one denomination have positive attitudes toward members of other religious groups. Denominational loyalties serve as poor predictors of attitudes or behavior.

Another restructuring is the proliferation of special-purpose groups. Special-purpose religious lobby or service groups have grown phenomenally since World War II, and most of them are free of denominational sponsorship. Indeed, denominations have become a sort of "holding [company] for an increasing number of special interest groups." Clearly, "the 7 percent that claims involvement in world hunger ministries is comparable to the 7 percent in the nation's largest Protestant denomination." But factor analysis uncovers two very divergent clusters of special-purpose groups, showing a divide between liberal and conservative religion in America.

Third is this religious divide. American religion has restructured from the consensus views of the 1950s into division along liberal or conservative lines, with each position of relatively equal strength—43 percent liberal versus 41 percent conservative. Each links to two quite distinctive understandings of civil religion.

Fourth, religion must now vie with a triad of secular values—individual freedom, material success, and the wonders of technology—as competing systems of legitimacy. Even opponents of these three are seduced into appealing to them.

Fifth, the growing role of the state in organizing modern society helps to explain the increasingly problematic symbolic boundaries between church and state.

I am not sure that Wuthnow's account does justice to what Wade Clark Roof and William McKinney have called "the fragmented mainline" Protestant churches. He skirts too much the paradox that while, as a statistical aggregate, so-called liberal religion is growing—perhaps within Catholicism and evangelicalism as much as anywhere else—the classic liberal churches have suffered massive membership loss. We need to know better why this has occurred and its significance for the geography of American religion.

Wuthnow writes with grace, wearing his massive learning and factual arsenal lightly. He probes the wider cultural significance and impact of American religion on society and vice versa. I suspect this thoughtful and informed book is likely to be the best work in the sociology of religion to appear in 1988.

JOHN A. COLEMAN

Graduate Theological Union
Berkeley
California

ECONOMICS

ASCH, PETER. *Consumer Safety Regulation: Putting a Price on Life and Limb*. Pp. x, 172. New York: Oxford University Press, 1988. $24.95.

Few would deny that concerns about the safety and health of users of the multifarious products available in the United States are now very high on the nation's political, legal, and social agendas and have been for at least two decades. These concerns are due, in large part, to the interaction of two forces. On the one hand, as consumers have approached higher levels of material satisfaction, they have, quite rationally, actively sought to enhance their nonmaterial welfare—for example, health and safety—through both market and nonmarket mechanisms. One could argue that to the degree that market processes, which work automatically and subtly, have failed them, consumers inevitably have increasingly and successfully resorted to political means, which in our tripartite, democratic government are neither automatic nor subtle. On the other hand, even if political and economic power were not at issue—which, of course, they are—there is reputedly a very substantial price being paid in terms of sacrificed output, as conventionally measured, by the diversion of productive factors to alleviate

these concerns—a diversion that is manifested by a quite substantial reallocation of resources and, hence, by a widespread unsettling of the economic structure of the *status quo ante*. With so much at stake, it is only natural that the political and economic consequences of this intersection of forces, and of how the battle has been waged, have received ample, mostly partisan, attention and even scholarly study. If these were Peter Asch's focus in *Consumer Safety Regulation*, the odds would be strongly against discovering any new insights.

Fortunately, Asch, an economist at Rutgers University, has a different aim. As it happens, in their abstract analyses of the functioning of the market mechanism, microeconomists have, somewhat serendipitously, made significant, piecemeal contributions to our knowledge of the by-product provisioning by markets of health and safety—and their negatives—and, so, to the design of effective public policy regarding this provisioning. Recall that the market mechanism is a social device, in a world of scarcity, through which, in principle, the individual, autarkic consumer acquires goods, produced by self-centered, merely profit-maximizing producing units, hopefully to achieve maximum individual welfare, subject to constraints, and, so, maximum social welfare, subject to different constraints. One can look at this process as follows: in a context of pervasive uncertainty, consumers examine themselves; they send more or less appropriate signals to producing units—firms—who receive the signals, though not necessarily the same ones; these units then act upon the received signals in a given economic environment; and so forth in a fairly long series of interactions and feedback. By parsing the workings of the market mechanism into these, and other, constituent elements one receives a superficial impression of its inherent complexity. Asch collates the results of theoretical economic—and empirical economic-psychological—research on the whole interconnected chain of the market mechanism and applies these to the issues of consumer health and safety. Principally, however, he examines, and raises questions about, those elements listed previously. For example, what can, or do, consumers find when they examine themselves? Is what they find conditioned by other consumers, or by the producing units, or not? Is whatever they find accurately reflected in the signals they send? Is the signal sent by the collectivity of consumers internally consistent? Is the sent signal accurately received, loud and clear, by the producing units? These questions involve, in some sense, access to information and the ability to convey it. In addition, as Asch shows, answers to them have interesting and often surprising implications for how markets, for example, provide products in which health and safety factors are important but largely incidental.

Asch's volume is a most provocative and useful book, inevitably, given the intricacy and exploratory nature of the matters discussed, leaving many policy-related issues unresolved.

M. O. CLEMENT

Dartmouth College
Hanover
New Hampshire

BROAD, ROBIN. *Unequal Alliance: The World Bank, the International Monetary Fund, and the Philippines.* Pp. xxvii, 352. Berkeley: University of California Press, 1988. No price.

Development aid programs in the Philippines provide the basis for this critical study of the International Monetary Fund (IMF) and the World Bank. The book is based primarily on leaks of confidential documents and interviews. Although partly polemical, the book also presents an informative account of economic policy in the Philippines in the 1970s and 1980s.

Readers should not be turned off in chapter 2 either by the misstatement that originally the IMF was not a macroeconomic-policy institution or by the incomplete quotation from Keynes referring to Bretton Woods as a monkey house. Throughout the book, Broad neglects the IMF's principal function as a regulator of exchange rates. Keynes's description of Bretton Woods as a monkey house reflects poorly on Keynes rather than on the U.S. delegation that Broad criticizes. Keynes was objecting to the U.S. policy of inviting a large number of delegates to the Bretton Woods conference from South America and other small countries.

This study attempts to show that the United States dominates economic policymaking in the Philippines, although this dominance is now exercised through the IMF and the World Bank. Broad criticizes the IMF and the World Bank for interfering secretly in domestic policymaking and for supporting export-oriented light manufacturing—apparel and electronics—rather than "genuine" industrialization, such as shipbuilding, textiles, and steel. It is claimed that export-oriented industrialization and its accompanying reduction in tariffs have decimated domestic enterprises, lowered the incomes of wage earners, resulted in industrial and financial concentration, strengthened state control of the economy, contributed to the country's large international debt, and, in addition, failed to achieve its growth objectives. Support for these conclusions is based primarily on comments by those involved rather than on statistics or analysis. Despite these limitations, Broad's numerous quotations provide information on public opinion not ordinarily found in economic or political studies.

Broad's view of the proper objectives of economic policy differs fundamentally from that of the IMF and the World Bank. To Broad, ideal conditions in less developed countries would replace the goal of economic growth with the goals of greater equality, wider participation in decision making, and less dependence on the world economy. The author concludes that the IMF and the World Bank have failed to adjust to the real conditions in the world in the 1980s—economic stagnation, declining world trade, curtailment of international lending, protectionism, technological unemployment, and increasing inter-national competition.

COLIN D. CAMPBELL
Dartmouth College
Hanover
New Hampshire

HEWETT, ED A. *Reforming the Soviet Econ-omy: Equality versus Efficiency.* Pp. xi, 404. Washington, DC: Brookings Institu-tion, 1988. $36.95. Paperbound, $16.95.

This appraisal of the prospects for Soviet economic reform is a remarkable piece of work. It is thorough in providing background and citing detailed developments. It is thought-ful in formulating a rigorous conceptual framework and applying relevant economic theory. It maintains a detached and objective stance. With all this, fortunately, it is gracefully written and therefore a pleasure to read.

The first four chapters set the stage. Hewett defines an economic reform, as opposed to a different set of policies, and explains how an economic reform effort can be launched, pursued, and eventually smothered by its opponents. He analyzes the strengths and weaknesses of the Soviet economic system, noting that the system provides security and equality much valued by the Soviet people. At the same time, the system is inefficient, costly, and technologically backward. The question is whether the weaknesses can be removed while the strengths are preserved. In two chapters he offers a sophisticated description of how the system is designed to operate, together with a solid analysis of how the system actually does operate. These chapters provide an up-to-date replacement for a great deal of the standard literature describing the Soviet economy's performance.

The next four chapters present a clinical examination of Soviet economic reform efforts since the death of Stalin. Important improve-ments were made by Khrushchev, and others were attempted by Kosygin and Brezhnev, but dissatisfaction with living standards and a slowdown in growth brought a national debate over proposed remedies. In a well-organized, incisive, 61-page analysis, Hewett offers a detailed review of the steps M. S. Gorbachev has followed from the spring of 1985 through the summer of 1987 in trying to start a basic reform of the whole economy. The book closes with a thoughtful chapter reflecting on the difficulties of measuring success or failure in these matters and on ways that the West might react to Soviet reform efforts.

Measurement difficulties arise because mere quantitative expansion is no longer enough; improvements are needed in the quality of Soviet goods and services and in the effectiveness with which Soviet economic insti-tutions perform—changes here are inherently hard to measure. The needed reforms are primarily internal ones on which the outside world can have little influence. The USSR would like, however, to export more manu-factured goods, but it can do so only if it meets quality standards imposed by the world market.

At the end of his book, Hewett offers a balanced, cautious discussion of Western interests in the outcome of Soviet reform efforts. Would success make the USSR more dangerous or more reasonable? His review of the factors to be considered will help every reader make judicious judgments as the Soviet reform process unfolds.

HOLLAND HUNTER
Haverford College
Pennsylvania

QUINN, DENNIS PATRICK, Jr. *Restructuring the Automobile Industry: A Study of Firms and States in Modern Capitalism.* Pp. xv, 395. New York: Columbia Univer-sity Press, 1988. $40.00.

The world automobile industry has changed rapidly in the last fifty years with increased diversification in location and continued ex-pansion in capacity. This led to a restructuring of the industry in the United States, Great Britain, and other industrial democracies. Quinn is interested in the role of the state in modern capitalism, specifically its role in the restructuring of this industry in the 1980s. After World War II, Western governments became welfare states concerned with economic growth and public welfare. Quinn examines the conditions under which market societies exhibit similar patterns of public policy toward private firms, given the interests of the welfare state.

The book has six well-written chapters with considerable data on the industry presented in interesting tables and figures, which clearly bring out the changing patterns in the industry. Chapter 1 is an overview of the book. This is followed, in chapter 2, by a brief economic and political history of the world auto industry since its inception. Chapter 3 is a description of the prosperity and decline of the American and British auto industries in the 1970s. Chapters 4 and 5 examine the policy similarities and the explanations for the similarities, respectively. The last chapter brings together the argument and presents its implications for private interests and the state.

Industry and government interests are interrelated in capitalist societies. In the case of the automobile industry, the commonality of interests between a government and domestic firms is substantial. In times of high auto production, firm and state interests are congruent. In a declining auto market, however, state interests and firm interests diverge in that the firms want to curtail production and close plants, while the government wants to keep up a high level of employment and avoid plant closings. Thus, under circumstances of economic decline, governments have adopted roughly similar policies. Although the British and American auto industries were quite different in size, products, and financial conditions, there were policy similarities in a period of economic decline that consisted in altering the cost structure of production to produce a commercially viable industry. Quinn extends the argument by using the corporate tax returns in four industrialized economies and finds similarities in public policy toward private firms in the United States, West Germany, Japan, and Britain. All four have subsidized their private corpora-

tions during periods of economic downturn. Quinn's hypothesis is that this can be accounted for by the welfare "state interest" explanation.

Although the book is more than just a detailed industry study, we would need to see a number of other links before any generalizations could be made about the role of the state in modern capitalist societies. There is no doubt that the occurrence of the yearly summits of the seven industrialized nations means that there is less resistance to having similar policies—in fact, part of the purpose of the meetings is to evolve similar policies for the benefit of their economies. This book is a useful addition to the public debate about public policy in the Western democracies. It is useful for policymakers, political scientists, economists, and those interested in the auto industry.

ASHOK BHARGAVA

University of Wisconsin
Whitewater

TEMIN, PETER. *The Fall of the Bell System*. Pp. xviii, 378. New York: Cambridge University Press, 1987. No price.

This book provides a detailed, blow-by-blow account of the complicated, sometimes irrational, events leading to the breakup of the world's largest corporation and regulated monopoly firm in 1983. The narrative, written by Massachusetts Institute of Technology economics professor Peter Temin with the close assistance of Johns Hopkins University history professor Louis Galambos, draws from a comprehensive information base consisting of Bell System records, public records, and personal interviews. The fact that the book was written at the invitation of the American Telephone and Telegraph Company, which opened its records to Temin, does not appear to have interfered in any way with the objectivity of the analysis.

In one sense, the book seems tedious and overly precise, but this is an inevitable byproduct of Temin's exhaustive effort to provide a complete and accurate picture of the many relevant factors that eventually led to the dissolution of the Bell System. While the casual reader must exert considerable patience in dealing with this maze of historical detail in the first seven chapters, such patience is rewarded by the excellent summary and overview of policy implications provided in the next, and final, chapter.

Importantly, the book helps one to appreciate the extreme complexity and cumbersome nature of government policy toward big business in the United States. In many ways, the nation had a good thing going in the old Bell System, which might appropriately be termed a benevolent monopoly. Indeed, its breakup—at least in the short run—appears to have left the average residential consumer in a worsened position while helping large business customers. Meanwhile, Temin clearly demonstrates that the goal of separating the competitive from the monopoly components of the telecommunications industry was not achieved in the final outcome of the case. Moreover, ideological viewpoints are shown to have prevailed over technological change as the primary moving force behind the restructuring of the American telecommunications industry.

In sum, Peter Temin deserves credit for a well-researched and carefully written analysis of one of the major economic events of American industrial history, an event that provides important lessons to be learned in relationship to the continuing deregulation movement in certain other important industries.

BERNARD P. HERBER
University of Arizona
Tucson

WINFIELD, RICHARD DIEN. *The Just Economy.* Pp. 252. New York: Routledge & Kegan Paul, 1988. $35.00.

Where does justice lie within the operations of an economy? Whereas Plato and other philosophers have asked where justice lies in the state and have treated the economic sphere as subsidiary to the political, Richard Winfield seeks a justice intrinsic to the economy. To find this, he starts with Hegel's treatment of "civil society" in the *Philosophy of Right* and builds it into a generally glowing, sometimes problematic defense of the market as an arena for the exercise of human freedom.

His argument, put simply, is that the market provides a sphere of civil freedom beyond natural needs, beyond the family, and apart from any higher telos, where humans may exercise voluntarism and experience reciprocity. Regardless of whether the needs—or ends—that a market participant seeks to fulfill are branded authentic or inauthentic, natural or artificial, essential or nonessential, nevertheless they are of one's own choosing and "attainable only through action toward others entailing the concomitant fulfillment of their respective ends." Furthermore, the means for one to become a market participant—that is, a commodity owner—are similarly "personally selected earning activities in reciprocity with others." Finally, the market, Winfield argues, is the nonexclusive institution par excellence; not only is it open to all classes, but "the market leaves its members at liberty to join several classes at once."

In the later chapters, especially one entitled "The Normative Principles of Public Intervention in the Market," Winfield takes up some of the more problematic aspects of his theory: "unemployment and overproduction," "the problem of monopoly formation," and "the recurrence of economic inequality." He concludes that both monopoly formation and the reproduction of an underprivileged class will require "never ending" remedial measures, but he argues against a "public dole," classless society, or other solutions that may interfere with the golden operation of freedom and reciprocity in the marketplace. What society can best do for the poor, Winfield suggests, is to empower them to become market participants. He argues, "By failing to earn the conventional livelihood to which all market participants are entitled, the victims of poverty are deprived not just of goods, but of the opportunity to exercise their autonomy as a rightful member of the community of interest."

Along the way to his own theory, Winfield rejects those of a number of others including Hobbes, Locke, Rousseau, and, most pointedly, Marx. He frequently belittles Marx's reasoning, calling Marx "confused" for claiming that market relations in themselves could involve any exploitation whatsoever.

While this, of course, may be all quite acceptable to conservatives, much of what Winfield has to say would have been rejected as apologetics for capitalism by even moderately leftist intellectuals—until a few years ago, at least. What makes the book more interesting today is that socialist thinkers have been trying to study just what might be valuable about markets, and to see if this is separable from class exploitation. Witness, for example, the market reforms in the Soviet Union and China. Moreover, Winfield, unlike many conservative economists, is well versed in Marx, even though his reading of the *1844 Manuscripts* and *Grundrisse* misses the dialectical richness of Marx's thought captured, for example, in the works of Martin Nicolaus and Richard J. Bernstein.

Unfortunately, Winfield's reading of Hegel is also a bit undialectical. By basing his theory wholly on Hegel's last and rather conservative work, *The Philosophy of Right*, and not putting this in the context of Hegel's earlier work—notably, *The Phenomenology of Mind*—Winfield misses the whole dimension of Hegel built on later by the left-Hegelians, such as Marcuse and the Frankfurt School. The crux here is that the moment of freedom to be experienced in the market is, in Hegelian dialectical terms, only that: a moment, or partial perspective. The next moment reveals that this freedom was also conditioned, by structures of all kinds: advertising institutions, governments, media. The legacy of Hegel hardly leads to the isolated adulation of the market that Winfield's book might imply.

Nevertheless, *The Just Economy* is a welcome addition to the growing interdisciplinary conversations that are helping to bring philosophy out of its analytic corner into an important societal role.

FRANCE H. CONROY
Burlington County College
Pemberton
New Jersey

OTHER BOOKS

ABBOTT, PHILIP. *Seeking Many Inventions: The Idea of Community in America.* Pp. x, 214. Knoxville: University of Tennessee Press, 1987. $19.95.

ANCKAR, DAG and ERKKI BERNDTSON, eds. *Political Science between the Past and the Future.* Pp. 132. Helsinki: Finnish Political Science Association, 1988. Paperbound, no price.

ANGLADE, CHRISTIAN and CHRISTIAN FORTIN. *The State and Capital Accumulation in Latin America.* Pp. xiii, 254. Pittsburgh, PA: University of Pittsburgh Press, 1985. $23.95.

BALL, RICHARD A. et al. *House Arrest and Correctional Policy: Doing Time at Home.* Pp. 180. Newbury Park, CA: Sage, 1988. Paperbound, $14.95.

BASSFORD, CHRISTOPHER. *The Spit-Shine Syndrome: Organizational Irrationality in the American Field Army.* Pp. xvii, 171. Westport, CT: Greenwood Press, 1988. $37.95.

BAUGHMAN, JAMES L. *Henry R. Luce and the Rise of the American News Media.* Pp. x, 264. Boston: G. K. Hall, 1987. $24.95.

BERNICK, MICHAEL. *Urban Illusions: New Approaches to Inner City Unemployment.* Pp. x, 243. Westport, CT: Praeger, 1987. $37.95.

BERNSTEIN, GAIL LEE and HARUHIRO FUKUI, eds. *Japan and the World: Essays in Japanese History and Politics.* Pp. xxii, 294. New York: St. Martin's Press, 1988. $55.00.

BLACK, ANTONY. *State, Community and Human Desire: A Group-Centered Account of Political Values.* Pp. xiii, 214. New York: St. Martin's Press, 1988. $35.00.

BONNER, JEFFREY P. *Land Consolidation and Economic Development in India.* Pp. x, 167. Riverdale, MD: Riverdale, 1987. $24.00.

BOWER, ROBERT T. *The Changing Television Audience in America.* Pp. ix, 172.

New York: Columbia University Press, 1988. $25.00.

BRENNER, REUVEN. *Batting on Ideas: Wars, Invention, Inflation.* Pp. x, 247. Chicago: University of Chicago Press, 1986. $32.00.

BRIMELOW, PETER. *The Patriot Game: National Dreams and Political Realities.* Pp. 310. Stanford, CA: Hoover Institution Press, 1986. $26.95.

CANCIAN, FRANCESCA M. *Love in America: Gender and Self-Development.* Pp. ix, 210. New York: Cambridge University Press, 1987. No price.

CHISMAN, FORREST and ALAN PIFER. *Government for the People: The Federal Social Role.* Pp. 316. New York: Norton, 1987. $17.95.

CONNOLLY, WILLIAM E. *Political Theory and Modernity.* Pp. xi, 196. New York: Basil Blackwell, 1988. $29.95.

CURRIE, DAVID. *The Constitution and the Supreme Court: The First Hundred Years, 1789-1888.* Pp. xiii, 504. Chicago: University of Chicago Press, 1986. $55.00.

DAHLBERG, KENNETH A. *New Directions for Agriculture and Agricultural Research.* Pp. xi, 436. Totowa, NJ: Rowman & Allanheld, 1986. $45.00. Paperbound, $18.95.

DALTON, RUSSELL J. *Citizen Politics in Western Democracies: Public Opinion and Political Parties in the United States, Great Britain, West Germany, and France.* Pp. xvi, 270. Chatham, NJ: Chatham House, 1988. Paperbound, $14.95.

DeLEON, DAVID. *Everything Is Changing: Contemporary U.S. Movements in Historical Perspective.* Pp. xvii, 285. New York: Praeger, 1988. $47.95. Paperbound, $14.95.

DIAMOND, EDWIN and STEPHEN BATES. *The Spot: The Rise of Political Advertising.* Revised ed. Pp. xiv, 425. Cambridge: MIT Press, 1988. Paperbound, $10.95.

DONEGAN, JANE B. *"Hydropathic Highway to Health": Women and Watercure in Antebellum America.* Pp. xx, 229. Westport, CT: Greenwood Press, 1986. $35.00.

DREIJMANIS, JOHN. *The Role of the South African Government in Tertiary Education.* Pp. xiii, 156. Johannesburg: South African Institute of Race Relations, 1988. Paperbound, $14.00.

DUCAT, STEPHEN. *Taken In: American Gullibility and the Reagan Mythos.* Pp. v, 161. Tacoma, WA: Life Sciences Press, 1988. Paperbound, $12.95.

DUCHACEK, IVO D. et al. *Perforated Sovereignties and International Relations: Trans-Sovereign Contacts of Subnational Governments.* Pp. xxii, 234. Westport, CT: Greenwood Press, 1988. $42.95.

FISHER, DAVID. *Morality and the Bomb: An Ethical Assessment of Nuclear Deterrence.* Pp. 133. New York: St. Martin's Press, 1985. $25.00.

FITZGERALD, RANDALL. *When Government Goes Private: Successful Alternatives to Public Services.* Pp. 330. New York: Universe, 1988. $24.95.

FRENKEL, JACOB A. *International Aspects of Fiscal Policies.* Pp. x, 408. Chicago: University of Chicago Press, 1988. $48.50.

GABRIEL, RICHARD A. *Military Incompetence: Why the American Military Doesn't Win.* Pp. xii, 207. New York: Hill & Wang, 1985. $16.95.

GALLAGHER, DOROTHY. *All the Right Enemies: The Life and Murder of Carlo Tresca.* Pp. xiii, 321. New Brunswick, NJ: Rutgers University Press, 1988. $24.95.

GILL, STEPHEN and DAVID LAW. *The Global Political Economy: Perspectives, Problems and Policies.* Pp. xxvi, 394. Baltimore, MD: Johns Hopkins University Press, 1988. $45.00. Paperbound, $17.50.

GIRARDET, EDWARD R. *Afghanistan: The Soviet War.* Pp. 259. New York: St. Martin's Press, 1986. $19.95.

GOEL, M. LAL. *Political Science Research: A Methods Handbook.* Pp. xiv, 296. Ames: Iowa State University Press, 1988. $25.95. Paperbound, $15.95.

GOLDSTEIN, TOM. *The News at Any Cost: How Journalists Compromise Their Ethics to Shape the News.* Pp. 301. New York:

Simon & Schuster, 1985. Paperbound, $8.95.

GRASSO, JUNE M. *Truman's Two-China Policy.* Pp. viii, 207. Armonk, NY: M. E. Sharpe, 1987. $29.50.

GREENE, OWEN, IAN PERCIVAL, and IRENE RIDGE. *Nuclear Winter.* Pp. 216. New York: Basil Blackwell, 1985. $34.95. Paperbound, $9.95.

HALL, JOHN R. *Gone from the Promised Land: Jonestown in American Cultural History.* Pp. xx, 381. New Brunswick, NJ: Transaction Books, 1987. $29.95.

HANLEY, DAVID. *Keeping Left? Ceres and the French Socialist Party.* Pp. viii, 278. New York: St. Martin's Press, 1988. Paperbound, $19.95.

HARDING, HARRY. *China and Northeast Asia: The Political Dimension.* Pp. xviii, 82. Lanham, MD: University Press of America, 1988. Paperbound, $6.00.

HARRIS, DAVID. *Justifying State Welfare: The New Right versus the Old Left.* Pp. ix, 181. New York: Basil Blackwell, 1987. $39.95.

HENDRY, JOY. *Becoming Japanese: The World of the Pre-School Child.* Pp. 194. Honolulu: University of Hawaii Press, 1987. $18.00.

HEYMANN, PHILIP B. *The Politics of Public Management.* Pp. xv, 196. New Haven, CT: Yale University Press, 1988. $22.50.

HUGHES, DEAN et al. *Financing the Agricultural Sector: Future Challenges and Policy Alternatives.* Pp. xi, 256. Boulder, CO: Westview Press, 1986. $24.85.

HULT, KAREN M. *Agency Merger and Bureaucratic Redesign.* Pp. xii, 219. Pittsburgh, PA: University of Pittsburgh, 1987. $28.95.

HUNT, MICHAEL H. *Ideology and U.S. Foreign Policy.* Pp. xiv, 237. New Haven, CT: Yale University Press, 1988. Paperbound, $8.95.

JENSON, ROBERT W. *America's Theologian: A Recommendation of Jonathan Edwards.* Pp. xii, 224. New York: Oxford University Press, 1988. $26.00.

JOHARI, J. C. *Contemporary Political Theory.* Revised ed. Pp. xvi, 756. New York: Apt Books, 1988. $45.00.

JONES, CHARLES O., ed. *The Reagan Legacy: Promise and Performance.* Pp. x, 311. Chatham, NJ: Chatham House, 1988. $25.00. Paperbound, $14.95.

KEARNEY, ROBERT N. and BARBARA DIANE MILLER. *Internal Migration in Sri Lanka and Its Social Consequences.* Pp. xvi, 143. Boulder, CO: Westview Press, 1987. Paperbound, no price.

KIRK, RUSSELL. *The Wise Men Know What Wicked Things Are Written on the Sky.* Pp. 138. Washington, DC: Regnery Gateway, 1987. $17.95. Paperbound, $9.95.

KOZLOV, VIKTOR. *The Peoples of the Soviet Union.* Pp. xi, 262. Bloomington: Indiana University Press, 1988. $37.50.

KUTTNER, ROBERT. *The Life of the Party: Democratic Prospects in 1988 and Beyond.* Pp. xviii, 265. New York: Penguin Books, 1988. Paperbound, $7.95.

LAMB, BRIAN et al. *C-SPAN: America's Town Hall.* Pp. xx, 393. Washington, DC: Acropolis Books, 1988. $19.95.

LANE, JAN-ERIK and SVANTE O. ERSSON. *Politics and Society in Western Europe.* Pp. x, 369. Newbury Park, CA: Sage, 1987. Paperbound, $16.50.

LEVY, EMANUEL. *John Wayne: Prophet of the American Way of Life.* Pp. xix, 379. Metuchen, NJ: Scarecrow Press, 1988. No price.

LITAN, ROBERT E. and CLIFFORD WINSTON, eds. *Liability: Perspectives and Policy.* Pp. xii, 248. Washington, DC: Brookings Institution, 1988. $28.95. Paperbound, $10.95.

LIXL-PURCELL, ANDREAS, ed. *Women of Exile: German-Jewish Autobiographies since 1933.* Pp. x, 231. Westport, CT: Greenwood Press, 1988. $35.00.

MARCH, JAMES G. *Decisions and Organizations.* Pp. vi, 458. New York: Basil Blackwell, 1988. $75.00.

MARCH, KATHRYN and RACHELLE TAQQU. *Women's Informal Associations in Developing Countries: Catalysts for Change?* Pp. xiii, 154. Boulder, CO: West-

view Press, 1986. Paperbound, $15.95.

MASTERS, ANTHONY. *Literary Agents: The Novelist as Spy.* Pp. viii, 271. New York: Basil Blackwell, 1987. $19.95.

McBROOM, PATRICIA. *The Third Sex: The New Professional Woman.* Pp. 283. New York: William Morrow, 1986. $16.95.

MEEHAN, EUGENE J. *The Thinking Game: A Guide to Effective Study.* Pp. ix, 244. Chatham, NJ: Chatham House, 1988. Paperbound, $12.95.

MERRIMAN, DAVID. *The Control of Municipal Budgets: Toward the Effective Design and Expenditure Limitations.* Pp. x, 170. Westport, CT: Quorum Books, 1987. $37.95.

MILLS, PATRICIA JAGENTOWICZ. *Woman, Nature, and Psyche.* Pp. xx, 266. New Haven, CT: Yale University Press, 1987. $26.50.

MOORE, WINFRED B., JOSEPH F. TRIPP, and LYON G. TYLER, Jr., eds. *Developing Dixie: Modernization in a Traditional Society.* Pp. xxiii, 353. Westport, CT: Greenwood Press, 1988. $39.95.

MOSLEY, PAUL. *Foreign Aid: Its Defense and Reform.* Pp. xiv, 264. Lexington: University Press of Kentucky, 1987. $25.00.

OHTA, HIROSHI. *Spatial Price Theory of Imperfect Competition.* Pp. x, 247. College Station: Texas A&M University Press, 1988. $34.50.

PAARLBERG, DON. *Toward a Well-Fed World.* Pp. xviii, 270. Ames: Iowa State University Press, 1988. $24.95.

PAPPE, ILAN. *Britain and the Arab-Israeli Conflict, 1948-51.* Pp. xxi, 273. New York: St. Martin's Press, 1988. $49.95.

POWASKI, RONALD E. *March to Armageddon: The United States and the Nuclear Arms Race, 1939 to the Present.* Pp. 300. New York: Oxford University Press, 1987. $19.95.

ROBBINS, KEITH. *Nineteenth-Century Britain: Integration and Diversity.* Pp. 199. New York: Oxford University Press, 1988. $29.95.

ROBERTS, ADAM and BENEDICT KINGSBURY, eds. *United Nations, Divided World: The UN's Roles in International*

Relations. Pp. xii, 287. New York: Oxford University Press, 1988. $59.00.

ROGERS, JOHN D. *Crime, Justice and Society in Colonial Sri Lanka.* Pp. x, 271. Riverdale, MD: Riverdale, 1987. $37.00.

ROHR, JOHN A. *To Run a Constitution: The Legitimacy of the Administrative State.* Pp. xv, 272. Lawrence: University Press of Kansas, 1986. $24.95. Paperbound, $12.95.

ROWEN, HENRY S. and CHARLES WOLF, Jr., eds. *The Future of the Soviet Empire.* Pp. xx, 368. New York: St. Martin's Press, 1988. $37.50. Paperbound, $14.95.

RUBIN, MICHAEL ROGERS and MARY TAYLOR HUBER. *The Knowledge Industry in the United States, 1960-1980.* Pp. xvi, 213. Princeton, NJ: Princeton University Press, 1986. No price.

RUESCHEMEYER, DIETRICH. *Power and the Division of Labor.* Pp. viii, 260. Stanford, CA: Stanford University Press, 1986. $32.50. Paperbound, $11.95.

SCHWARTZMAN, DAVID. *Games of Chicken: Four Decades of U.S. Nuclear Policy.* Pp. xi, 233. New York: Praeger, 1988. $37.95.

SHARPES, DONALD K. *Curriculum Traditions and Practices.* Pp. 121. New York: St. Martin's Press, 1988. $39.95.

SONDERSON, STEVEN E. *The Transformation of Mexican Agriculture: International Structure and Politics of Rural Change.* Pp. xxii, 324. Princeton, NJ: Princeton University Press, 1986. $42.00. Paperbound, $10.95.

STEETEN, PAUL, ed. *Beyond Adjustment: The Asian Experience.* Pp. xii, 274. Washington, DC: International Monetary Fund, 1988. Paperbound, no price.

TAYLOR, PAUL and AJR GROOM, eds. *International Institutions at Work.* Pp. vii, 245. New York: St. Martin's Press, 1988. $39.95.

TESH, SYLVIA NOBLE. *Hidden Arguments: Political Ideology and Disease Prevention Policy.* Pp. viii, 215. New Brunswick, NJ: Rutgers University Press, 1988. $30.00.

TOCQUEVILLE, ALEXIS DE. *Journeys to England and Ireland.* Edited by J. P. Mayer. Pp. 255. New Brunswick, NJ: Transaction Books, 1988. Paperbound, $19.95.

UTTON, M. A. *The Economics of Regulating Industry.* Pp. x, 243. New York: Basil Blackwell, 1986. No price.

WATSON, SOPHIE and HELEN AUSTERBERRY. *Housing and Homelessness: A Feminist Perspective.* Pp. 186. Boston: Routledge & Kegan Paul, 1986. Paperbound, $16.95.

WEHR, DEMARIS S. *Jung Feminism: Liberating Archetypes.* Pp. xii, 148. Boston: Beacon Press, 1987. $17.95.

WEISS, THOMAS G. *Multilateral Development Diplomacy in UNCTAD.* Pp. xviii, 187. New York: St. Martin's Press, 1986. $30.00.

WELLS, CLARE. *The UN, Unesco and the Politics of Knowledge.* Pp. xviii, 281. New York: St. Martin's Press, 1987. $32.50.

WHITE, MERRY. *The Japanese Educational Challenge: A Commitment to Children.* Pp. x, 210. New York: Free Press, 1987. $18.95.

WHITE, STEPHEN and ALEX PRAVDA, eds. *Ideology and Soviet Politics.* Pp. viii, 258. New York: St. Martin's Press, 1988. $49.95.

WILLIAMS, HOWARD. *Concepts of Ideology.* Pp. xiii, 136. New York: St. Martin's Press, 1988. $35.00.

WILLIAMS, ROBERT C. *The Other Bolsheviks: Lenin and His Critics, 1904-1914.* Pp. vii, 233. Bloomington: Indiana University Press, 1986. $27.50.

WINKLE, KENNETH J. *The Politics of Community: Migration and Politics in Antebellum Ohio.* Pp. xiii, 239. New York: Cambridge University Press, 1988. No price.

WOLF, ROBERT PAUL. *Moneybags Must Be So Lucky: On the Literary Structure of Capital.* Pp. 82. Amherst: University of Massachusetts Press, 1988. $20.00. Paperbound, $8.95.

ZAWODNY, J. K. *Death in the Forest: The Story of the Katyn Forest Massacre.*

Pp. xv, 235. New York: Hippocrene Books, 1988. Paperbound, no price.

ZEITLIN, MAURICE. *The Civil Wars in Chile, Or the Bourgeois Revolutions That Never Were.* Pp. xiii, 265. Princeton, NJ: Princeton University Press, 1988. Paperbound, $14.95.

ZUCKER, NORMAN L. and NAOMI FLINK ZUCKER. *The Guarded Gate: The Reality of American Refugee Policy.* Pp. xx, 342. New York: Harcourt Brace Jovanovich, 1987. $22.95.

ZWIRN, JEROLD. *Congressional Publications and Proceedings.* 2d ed. Pp. xvii, 299. Englewood, CO: Libraries Unlimited, 1988. $27.50.

Abrams, Richard, 10
ABRAMS, RICHARD M., The U.S. Military and
 Higher Education: A Brief History, 15-28
Academic freedom, 19
Air Force Office of Scientific Research, 32, 43, 60,
 66, 84, 97
Antiballistic missiles, 45-46
Antinuclear protest, 126-27
Applied Physics Laboratory, 99, 120-29
Army Research Office, 32, 43, 60, 66, 84, 97
Army Specialized Training Programs, 96
Arroyo Center, 74-75
Atomic Energy Commission, 32

Baranger, Elizabeth, 10
BARANGER, ELIZABETH UREY, see GER-
 JUOY, EDWARD, coauthor
California Institute of Technology (Cal Tech), 69,
 74-75, 80
CAN UNIVERSITIES COOPERATE WITH
 THE DEFENSE ESTABLISHMENT? Carl
 Kaysen, 29-39
Carnegie-Mellon University, 75-76
Carter administration, 135
Charles Stark Draper Laboratory, 74
Classified research, 21-22, 23, 32, 35, 50-51, 55, 72,
 74, 123, 126, 127, 128, 134, 135, 137-38, 148-49
 see also Information transfer
Cold War, 21-22
Computer science, research in, 82-93, 133
Conscription, 112-16, 118-19
 see also Selective Service
CONSEQUENTIAL CONTROVERSIES, David A.
 Wilson, 40-57
Cooper, Julian, 13
COOPER, JULIAN, The Military and Higher
 Education in the USSR, 108-19

Defense Academic Research Projects Agency, 43
Defense Academic Research Support Program, 10
Defense Advanced Language and Area Studies
 Program, 106
Defense, Department of
 opposition to Defense Department support of
  research, 73-75, 78-79, 126
 skewing of science and research, 53-54, 71, 77, 78,
  146-48, 150-51
 support of research, 23-25, 26, 27, 36, 61, 66-73, 77,
  78-79, 86, 89, 92-93, 97-100, 133, 138-39,
  141-54
Defense Language Institute, 103
DeLauer, Richard, 13

DeLAUER, RICHARD D., The Good of It and Its
 Problems, 130-40
DoD, SOCIAL SCIENCE, AND INTERNA-
 TIONAL STUDIES, Richard D. Lambert,
 94-107
DoD-University Forum, 136

ELECTRONICS AND COMPUTING, Leo Young,
 82-93
Electronics, research in, 82-93
Export controls, 149

Federally Funded Research and Development
 Centers (FFRDCs), 33-34, 60-61, 69-70, 74,
 80, 99, 132
Foreign language education, and the Defense
 Department, 102

General Defense Intelligence Program, 105
Gerjuoy, Edward, 10
GERJUOY, EDWARD and ELIZABETH UREY
 BARANGER, The Physical Sciences and
 Mathematics, 58-81
GOOD OF IT AND ITS PROBLEMS, THE,
 Richard D. DeLauer, 130-40
Graduate students, 56, 62-63, 69, 80-81, 135, 147,
 149-50

Higher Education Act, 105

Information transfer, 55-56, 72-73, 76, 79
 see also Classified research; Technology transfer
International relations and strategic studies, relationship with the Defense Department, 101-2

Jet Propulsion Laboratory (JPL), 69, 120-29
Johnson, Lyndon B., 36
Joint Services Electronics Program (JSEP), 86-88

Kaysen, Carl, 10
KAYSEN, CARL, Can Universities Cooperate
 with the Defense Establishment? 29-39
Killian, James R., Jr., 21-22
Kistiakowsky, Vera, 14
KISTIAKOWSKY, VERA, Military Funding of
 University Research, 141-54
Korean War, 98, 143

Lambert, Richard, 13
LAMBERT, RICHARD D., DoD, Social Science,
 and International Studies, 94-107

Language and area studies, relationship with the Defense Department, 103-7
Los Alamos National Laboratory, 33

Mansfield Amendment, 27, 36, 51-52, 66-67, 77-78, 100, 139, 143
Massachusetts Institute of Technology (MIT), 10, 21-22, 69, 74, 151, 152-53
Mathematics, research in, 58-81
McCarthyism, 34-35, 98
MILITARY AND HIGHER EDUCATION IN THE USSR, THE, Julian Cooper, 108-19
MILITARY FUNDING OF UNIVERSITY RESEARCH, Vera Kistiakowsky, 141-54
Morrill Land-Grant College Act (1862), 9, 16, 18, 20
Muller, Steven, 13
MULLER, STEVEN, The View of the Big Performers, 120-29

National Academy of Science, 133-34
National Aeronautics and Space Administration (NASA), funding of research, 66
National Defense Act (1916), 19
National Defense Act (1920), 19
National Defense Education Act, 103, 105
National Defense Foundation (NSF), 32, 60, 61, 66, 78, 86, 97, 102, 143
National Security Agency, and support of foreign language instruction, 102-3
Nixon administration, 135
Nixon, Richard M., 36-37

Office of Naval Research, 31-32, 41, 59-60, 66, 84, 97, 143

Pacifists, 118
Peer review, 21, 23, 32, 54-55
PHYSICAL SCIENCES AND MATHEMATICS, THE, Edward Gerjuoy and Elizabeth Urey Baranger, 58-81
Physical sciences, research in, 58-81
Project Camelot, 51, 100
Public service, 16-17, 19, 122, 128, 146, 151

Reagan administration, 24, 74
Reagan, Ronald, 37
Research grants, university dependence on, 80-81
Reserve Officers' Training Corps (ROTC), 19, 20, 25, 35, 49, 97
Rockefeller, Nelson, 134

Selective service, 49
 see also Conscription
Social responsibility, 48, 151, 153
Social science, and relationship with the Defense Department, 99-101
Software Engineering Institute, 75-76
Stanford Linear Accelerator Center Synchrotron Radiation Facility, 152
Stanford Research Institute, 74
Strategic Bombing Survey, 96
Strategic Defense Initiative, 28, 37-38, 46-47, 74, 127, 144-145, 151, 152-53

Tartu University, 116
Technology transfer, 90-91
 see also Information transfer
Technology transition, 91
Tomsk, boycott of military training in, 116

University of California, 33
University Research Initiative, 24, 88, 99, 100
U.S. MILITARY AND HIGHER EDUCATION: A BRIEF HISTORY, THE, Richard M. Abrams, 15-28
USSR
 civil defense, 111-12
 military faculties and departments in higher education, 111, 116-17
 military-patriotic education, 112
 relationship between the military and higher education, 108-19
 teachers from industry in higher education, 111

V-12 programs, 96
Vietnam war, 35-36, 47, 74, 85, 98, 127, 133-34, 143
VIEW OF THE BIG PERFORMERS, THE, Steven Muller, 120-29

"War research," 47-48, 50-51
Weapons technology, 97
WILSON, DAVID A., Consequential Controversies, 40-57
WILSON, DAVID A., Preface, 9-14
World War I, 95
World War II, 20, 31, 83, 95-96, 103, 122, 131

Young, Leo, 10
YOUNG, LEO, Electronics and Computing, 82-93